Biodynamic Farming
An Ancient Agricultural Wisdom

CONTENT VIBES
PUBLISHERS ▪ PRINTERS ▪ AGGREGATORS

About the Authors

Dr. M. Ramanjineyulu Ph.D., (Agronomy) works as Assistant Professor in Sri Krishnadevaraya college of Horticultural Sciences, Ananthapuramu, Andhra Pradesh, India. He holds three years of teaching and research experience. He has to his credit around 9 scientific research publications in national and international journals besides several popular articles. He is the author of several book chapters on different aspects of agronomy. He was awarded with the Best Thesis Award for his excellent work in Ph.d and awarded Young Agronomist Award for his research in various aspects of agronomy.

Dr. Mude Ashok Naik has published more than 20 papers comprising of research articles, popular articles and book chapters. He received the Provisional Fellowship Award from Union Grants Commission to pursue Ph.D and was recipient of the Professor MS Swaminathan Best Young Investigator Award, India Star Republic Award, Research Excellence Award for his research works and Best Ph.D Thesis Award. He holds teaching and research experience, and specializes in seed production programs in various agriculture and horticultural crops.

Dr. S.N. Abhilash Naik, holds an PhD. (Ag.) in Agronomy from Agricultural College, Bapatla, ANGRAU. With a strong academic background in agronomic sciences, he has contributed more than 10 research articles, as well as book chapters published in NAAS-indexed platforms. He holds teaching and research experience, with research interests spanning soil health, organic farming techniques and sustainable agriculture. He has extensive experience in paddy seed production and other agriculture crops. He also specializes in integrated farming systems and maintenance.

Dr. A. Bharathi works as Assistant Professor of Agronomy at Kalasalingam Academy of Research and Education, Srivilliputtur, Tamil Nadu. She completed her academics from Tamil Nadu Agricultural University, Coimbatore. She has published more than 10 research papers and more than 20 popular articles in reputed journals. She was awarded the Best Outgoing Student and Best Thesis Award for her Doctoral Program. She specializes in evaluation of different physical and chemical problems and has expertise in crop modeling with hands-on training with different crop simulation models.

Mr. Singireddy Prabhu Mitra Reddy holds an M.Sc. in Agronomy from Nagaland University and a B.Sc. in Agriculture from Tamil Nadu Agricultural University. With hands-on experience in crop production, soil analysis, and sustainable farming techniques, he has contributed to research on traditional rice varieties and integrated nitrogen management in sesame. His professional background includes work at Krishi Vigyan Kendra, focusing on climate-resilient agriculture and farmer training. As a co-author of this book on Biodynamic Farming, he brings expertise in organic cultivation, natural resource management, and innovative agricultural practices.

Ms. Sibbala Yoshitha holds an M.Sc. (Ag.) in Agronomy from Centurion University of Technology and Management, Paralakhemundi, Odisha. She works as Assistant Professor in Sri Krishnadevaraya College of Agricultural sciences, Ananthapuramu, Andhra Pradesh. With a strong academic background in agronomic sciences, Yoshitha has contributed to the field through research articles as well as book chapters published in NAAS-indexed platforms. Passionate about sustainable agricultural practices. Her research interests include soil health, organic farming techniques, and holistic crop management.

Ms. Deepika J. graduated from University of Agricultural Sciences, Raichur with first class. She obtained her M.Sc. (Ag). from Lovely Professional University, Punjab with first class. She has published 7 papers in national and international journals, comprising research and popular articles and book chapters. She has participated in national and international seminars, conferences and workshops, and was also awarded the Best PG thesis award. She holds experience in research on agroforestry-based faming systems and has extensive knowledge in agro-forestry nurseries.

Dr. Marati Sainath Rao is an Assistant Professor in the Department of Spices, Plantation, Medicinal and Aromatic Plants at Sri Krishnadevaraya College of Horticultural Sciences, Anantapur. With a Master's in Medicinal and Aromatic Plants from HNB Garhwal University, he has also been a recipient of the 'Young Horticulturist Award'. He has contributed significantly to agricultural literature through numerous research papers and articles in prestigious journals. His practical experience includes work with the Uttarakhand Biodiversity Board and extensive field research in plant conservation.

Biodynamic Farming
An Ancient Agricultural Wisdom

M. Ramanjineyulu
Mude Ashok Naik
S.N. Abhilash Naik
A. Bharathi
Singireddy Prabhu Mitra Reddy
Sibbala Yoshitha
Deepika J.
Marati Sainath Rao

CRC Press
Taylor & Francis Group
Boca Raton London New York

CRC Press is an imprint of the
Taylor & Francis Group, an **informa** business

CONTENT VIBES
PUBLISHERS ▪ PRINTERS ▪ AGGREGATORS

First published 2027
by CRC Press
4 Park Square, Milton Park, Abingdon, Oxon, OX14 4RN

and by CRC Press
2385 NW Executive Center Drive, Suite 320, Boca Raton FL 33431

CRC Press is an imprint of Informa UK Limited

British Library Cataloguing-in-Publication Data
A catalogue record for this book is available from the British Library

Print edition not for sale in South Asia (Bangladesh, Bhutan. India, Nepal, Pakistan and Sri Lanka).

ISBN: 9781041349105 (hbk)
ISBN: 9781041389330 (pbk)
ISBN: 9781042015597 (ebk)

DOI: 10.1201/9781042015597

Typeset in Times New Roman
by Content Vibes

CONTENT VIBES
PUBLISHERS • PRINTERS • AGGREGATORS

Foreword

As agriculture faces increasing challenges such as soil degradation, water scarcity, biodiversity loss and the growing demand for sustainable food production - Biodynamic farming offers a time-tested, holistic solution. This ancient yet progressive agricultural approach harmonizes the rhythms of nature with the cultivation of crops and livestock, fostering ecosystems that are not only productive but also resilient and regenerative. By emphasizing soil health, biodiversity and cosmic influences, Biodynamic farming provides a comprehensive framework that addresses both environmental sustainability and food security.

This book, Biodynamic Farming: An Ancient Agricultural Wisdom, fills a critical gap in the agricultural literature. While there is abundant information on conventional and organic farming, few resources focus specifically on the principles and practices of biodynamics, making this work especially valuable. The authors have chosen a subject that is both historically rich and increasingly relevant to the future of farming. Their thorough research and thoughtful analysis present biodynamic farming as a vital and transformative approach to agriculture, blending ecological balance with spiritual insight.

Written in clear, accessible language, this book serves as a practical guide for a diverse audience. Whether you are an experienced farmer, a student of agriculture, a horticulturist, or someone interested in sustainable practices, the content is designed to make complex concepts easy to understand and apply. It covers the foundation principles of biodynamic farming as well as practical aspects, including soil fertility, planting calendars, biodynamic preparations and livestock integration. The book also includes carefully selected images and illustrations, which provide a visual context that enhances the reader's understanding of these topics.

With the growing global focus on sustainable farming practices, this book will be an invaluable resource for anyone looking to adopt or deepen their knowledge of biodynamic farming. As an Agricultural scientist and professor of Agronomy, I firmly believe that biodynamic farming offers a transformative approach that can revitalize Agri & Horticultural ecosystems while ensuring food production is in harmony with nature.

I highly recommend this book to farmers, educators, researchers and anyone passionate about advancing sustainable and biodynamic farming practices. It is an informative, practical and inspirational guide, one that will equip you with the knowledge to navigate the challenges and opportunities of modern agriculture.

Mallu Srinivasa Reddy

Preface

In an era where industrial agriculture dominates our food systems and environmental concerns reach critical levels, the ancient wisdom of biodynamic farming emerges as a beacon of hope and sustainability. Biodynamic farming, first introduced by Rudolf Steiner in 1924, represents far more than just another agricultural method. It embodies a profound understanding of the intricate relationships between soil, plants, animals, and cosmic forces. As we grapple with climate change, soil degradation, and biodiversity loss in the 2020s, the principles of biodynamic agriculture offer solutions that transcend conventional farming paradigms.

This book serves as both an introduction and a detailed exploration of biodynamic farming practices, carefully crafted to bridge the gap between ancient wisdom and modern agricultural needs. Unlike many technical agricultural texts, we have deliberately maintained an accessible writing style, ensuring that farmers, students, researchers, and enthusiasts alike can grasp these complex concepts with ease. For practicing farmers, this guide provides practical insights into implementing biodynamic preparations, understanding planting calendars, and integrating livestock into farming systems. Students will find the detailed examination of soil health dynamics, lunar influences, and the systematic approach to sustainable agriculture that biodynamic farming offers. The chapters progress logically from basic concepts to advanced applications, including certification standards and economic considerations that are crucial for commercial success.

The timing of this publication coincides with a growing global awareness of the need for sustainable agricultural practices. As consumers become increasingly conscious of food quality and environmental impact, biodynamic farming offers a proven pathway to producing nutrient-rich foods while regenerating ecosystems. This book serves as a practical roadmap for those seeking to embrace these methods, whether on a small homestead or a large commercial operation. As we look toward the future of agriculture, the principles outlined in this book become increasingly relevant. This book not only teaches the "how" of biodynamic farming but also explains the "why," empowering readers to adapt these principles to their unique circumstances.

It is our sincere hope that this book will serve as both a practical guide and a source of inspiration for the next generation of farmers, researchers, and agricultural professionals. In these pages, you will find the tools and knowledge needed to participate in the transformation of our agricultural systems toward greater sustainability, resilience, and harmony with nature.

Authors

Contents

1

Introduction, History, National and International Status of Biodynamic Farming

Abstract

Biodynamic farming, a holistic agricultural system developed in the early 20th century, integrates ecological and spiritual principles to enhance soil fertility and plant health. Originating from the ideas of Rudolf Steiner, biodynamic practices emphasize the interconnection between soil, plants, animals, and cosmic rhythms. This article provides a comprehensive overview of biodynamic farming, tracing its historical roots, current national and international status, and its impact on sustainable agriculture. Key elements such as composting, crop rotation, and lunar planting cycles are explored. The global reach and recognition of biodynamic certification, its role in contemporary organic farming, and the challenges it faces are also discussed. By examining biodynamic farming's past and present, this article aims to highlight its potential to contribute to sustainable and regenerative agricultural practices worldwide.

Introduction

Biodynamic farming is a holistic and sustainable approach to agriculture that goes beyond organic practices, incorporating spiritual, ethical, and ecological dimensions into farming and food production. Rooted in the early 20th century, this method is based on the teachings of Rudolf Steiner, an Austrian philosopher, who laid the groundwork for what has become a global movement aimed at creating a harmonious balance between nature and agriculture. Unlike conventional farming, which often relies heavily on chemical inputs and monoculture practices, biodynamic farming emphasizes the interdependence of all elements within a farming system. This includes the soil, plants, animals, and even cosmic forces. The approach seeks to enhance the vitality of the farm as a living organism, fostering a self-sustaining and self-regulating ecosystem.

Central to biodynamic farming is the concept of the farm as a closed-loop system, where all inputs are produced on-site. This includes composting

DOI: 10.1201/9781042015597-1

organic waste, rotating crops to maintain soil fertility, and using herbal and mineral preparations to enhance the health and resilience of the farm. These preparations are often made from specific plants, minerals, and animal manure, which are applied according to a biodynamic calendar that aligns farming activities with the lunar and cosmic cycles.

The appeal of biodynamic farming lies not only in its environmental benefits but also in its philosophical underpinnings, which promote a deep connection between the farmer, the land, and the cosmos. This connection is believed to produce food that is not only more nutritious but also imbued with a unique vitality that contributes to the well-being of those who consume it.

Over the years, biodynamic farming has evolved into a global movement, with certified farms and vineyards in many countries. It has gained recognition for its contributions to sustainable agriculture, soil health, and food quality, and has inspired a broader conversation about the future of farming in the face of environmental challenges.

1. History of Biodynamic Farming

The origins of biodynamic farming can be traced back to the early 20th century, during a time when industrial agricultural practices were beginning to take a toll on soil fertility and crop quality. In 1924, Austrian philosopher and scientist Rudolf Steiner delivered a series of eight groundbreaking lectures in Koberwitz, which is now known as Kobierzyce, Poland. These lectures, collectively referred to as the "Agriculture Course," addressed the pressing concerns of farmers who were witnessing the detrimental effects of chemical fertilizers and pesticides on their lands and produce.

Steiner's approach was revolutionary for his time. He proposed a holistic view of farming, one that recognized the farm as a self-sustaining organism where all components—soil, plants, animals, and even cosmic forces—are interconnected and mutually dependent. This idea marked the inception of biodynamic farming, which emphasizes working in harmony with natural rhythms and cycles. Steiner also introduced the concept of using specially prepared composts and natural remedies, known as biodynamic preparations, to enhance soil vitality and plant health.

The "Agriculture Course" laid the philosophical and practical foundation for biodynamic practices, merging scientific observations with spiritual insights. Steiner's teachings drew upon anthroposophy, a spiritual philosophy he developed, which seeks to understand the interrelationship between humans and the natural world. He emphasized the importance of viewing the farm not merely as a production unit but as a living, breathing entity.

Biodynamic farming gained traction in the decades that followed, with early practitioners refining and expanding upon Steiner's principles. Today, it is recognized as one of the earliest forms of organic farming, influencing sustainable agricultural movements worldwide. The integration of cosmic influences, crop rotation, and the avoidance of synthetic inputs continues to define biodynamic agriculture, making it a pioneering approach to regenerative farming practices.

1.1 Early Development and Spread

The origins of biodynamic agriculture trace back to the early 20th century when Rudolf Steiner delivered a series of lectures in 1924 in Koberwitz, Germany (now Kobierzyce, Poland). These lectures, collectively known as the "Agriculture Course," laid the foundation for the biodynamic farming movement. Responding to concerns about declining soil fertility and the detrimental effects of chemical fertilizers, Steiner emphasized a holistic and spiritual approach to farming that integrated cosmic and earthly influences.

Following these seminal lectures, a dedicated group of farmers and scientists came together to form the Experimental Circle of Anthroposophic Farmers and Gardeners. This group took on the task of implementing and refining Steiner's ideas through practical experimentation and collaboration. They focused on developing specific biodynamic preparations—herbal, mineral, and manure-based formulations—designed to enhance soil health, plant vitality, and ecological balance. These preparations were intended to work in harmony with natural rhythms, including lunar and planetary cycles.

The biodynamic movement quickly gained traction in Europe, with the first biodynamic farms emerging in Germany, Switzerland, and the Netherlands. Early adopters of this innovative method were drawn by the promise of revitalized soils and improved crop quality. They began experimenting with Steiner's prescribed practices, such as the use of compost inoculants and the application of biodynamic sprays. Over time, these practices were refined and adapted to suit diverse regional climates and agricultural conditions.

By the mid-20th century, the principles of biodynamic agriculture had spread beyond Europe, reaching North America, Australia, and parts of Asia and Africa. This global expansion was facilitated by a growing network of biodynamic organizations and certification bodies, such as Demeter International, which established rigorous standards for biodynamic farming. The movement's international growth was further supported by scientific studies, farmer testimonials, and an increasing public interest in sustainable and organic agricultural practices.

Today, biodynamic farming continues to influence modern agriculture, offering a pioneering model for regenerative and holistic farming systems. The foundational work of the Experimental Circle and the commitment of early biodynamic farmers remain central to the movement's enduring legacy.

2. Key Principles and Practices of Biodynamic Farming

Biodynamic farming is a holistic approach to agriculture that emphasizes harmony between nature, the cosmos, and human activities. It builds upon a set of key principles and practices that distinguish it from other forms of organic and conventional agriculture. These principles aim to create a self-sustaining and regenerative farming system that not only produces healthy food but also contributes to environmental balance and biodiversity. Here, we explore the essential components that define biodynamic farming:

2.1 The Farm as a Self-Sustaining Organism

Biodynamic farming treats the farm as a living, interconnected organism where each component—soil, plants, animals, and humans—plays a vital role. This principle encourages farms to minimize external inputs and focus on creating closed nutrient cycles. Livestock, for instance, provides manure that enriches the soil, while diverse plant species contribute to ecological balance and pest control.

2.2 Biodynamic Preparations

A hallmark of biodynamic agriculture is the use of specific preparations made from natural materials like herbs, minerals, and animal manure. These preparations are applied to the soil or compost to enhance microbial activity and improve soil fertility. Examples include Preparation 500 (horn manure) and Preparation 501 (horn silica), which are believed to strengthen plant health and soil vitality.

2.3 Composting

Composting is central to biodynamic farming, transforming organic waste into nutrient-rich humus that sustains soil health. Biodynamic composting involves the use of specialized preparations to accelerate decomposition and enhance the compost's effectiveness. Healthy compost ensures long-term fertility and resilience against soil degradation.

2.4 Crop Rotation and Diversification

To maintain soil health and reduce pest outbreaks, biodynamic farmers practice crop rotation and diversification. Rotating crops prevents nutrient depletion and disrupts pest cycles, while diversifying plant species fosters resilience against diseases and environmental stresses. This practice also supports a broader range of beneficial organisms.

2.5 Lunar and Cosmic Rhythms

Biodynamic farming incorporates the influences of lunar and cosmic rhythms on plant growth and soil health. Farmers follow a biodynamic calendar to schedule planting, harvesting, and other activities in harmony with celestial cycles. This practice is believed to optimize agricultural productivity and align farming practices with natural forces.

By integrating these principles, biodynamic farming offers a regenerative approach to agriculture that nurtures the environment while producing high-quality food. Its holistic philosophy underscores the interconnectedness of all living systems and encourages stewardship of the Earth for future generations.

3. National Status of Biodynamic Farming

Biodynamic farming has gained recognition and acceptance in various countries, each with its unique developments and contributions to the movement.

3.1 Biodynamic Farming in India

India's agricultural practices have long been rooted in traditional and sustainable methods, many of which align closely with the principles of biodynamic farming. This ancient agricultural philosophy, which emphasizes the interconnectedness of soil, plants, animals, and cosmic rhythms, has found a natural resonance within India's culturally diverse and ecologically rich landscape. Over the past few decades, biodynamic farming has gained significant traction in the country, offering a viable alternative to chemical-intensive conventional agriculture.

One of the key figures in promoting biodynamic farming in India is Peter Proctor, a New Zealand agriculturalist often referred to as the "father of modern biodynamic farming." In the 1990s, Proctor introduced biodynamic methods to Indian farmers, demonstrating their potential to rejuvenate degraded soils, increase crop resilience, and enhance nutritional quality. His efforts sparked a movement, leading to widespread awareness and adoption of these practices in various regions of the country.

The Indian Biodynamic Association, along with numerous non-governmental organizations (NGOs), has played a pivotal role in advancing biodynamic farming. These organizations provide training, resources, and support to farmers, enabling them to transition from conventional to biodynamic systems. Such initiatives are particularly impactful in areas facing severe challenges like soil degradation, water scarcity, and declining agricultural productivity.

For instance, biodynamic preparations such as compost enriched with specific herbs and the application of natural sprays have been successfully used to restore soil health and improve water retention in arid regions.

States like Maharashtra, Tamil Nadu, and Uttarakhand have emerged as hubs for biodynamic farming in India. Farmers in these areas have reported improved yields, better soil fertility, and reduced dependency on chemical inputs. Moreover, biodynamic farming has opened up new opportunities in organic and export markets, as biodynamic produce is highly valued for its quality and sustainability. Beyond economic benefits, biodynamic farming in India is also closely tied to the preservation of indigenous knowledge and biodiversity. Traditional Indian farming practices, such as crop rotation, intercropping, and the use of natural fertilizers, align seamlessly with biodynamic principles. This synergy has allowed Indian farmers to innovate and adapt biodynamic methods to local conditions, creating a sustainable agricultural model that respects both nature and community.

As the world increasingly recognizes the importance of sustainable agriculture in combating climate change and ensuring food security, India's growing biodynamic farming movement offers a beacon of hope. Through continued efforts by farmers, organizations, and policymakers, biodynamic farming has the potential to transform Indian agriculture, making it more resilient, eco-friendly, and economically viable.

1.2 International Status of Biodynamic Farming

Biodynamic farming has gained international recognition and continues to expand globally. The movement's principles and practices resonate with farmers and consumers seeking sustainable and environmentally friendly alternatives to industrial agriculture.

i. **Germany:** Germany, the birthplace of biodynamic farming, remains a stronghold for the practice. The country is home to Demeter, the world's oldest certification organization for biodynamic agriculture. Founded in 1928, Demeter has played a crucial role in promoting biodynamic standards and supporting farmers through education and certification programs. Today, Germany boasts a significant number of biodynamic farms, producing a wide range of crops, dairy products, and wines.

ii. **United States:** In the United States, biodynamic farming has seen steady growth, particularly in the wine industry. Biodynamic vineyards in California, Oregon, and Washington are renowned for their high-quality wines produced using biodynamic principles. Organizations like the Biodynamic Association and Demeter USA have been instrumental in advancing the movement by providing resources, training, and certification to American farmers.

iii. **Australia:** Australia is another country where biodynamic farming has established a strong presence. The Australian Demeter Biodynamic Certification (Demeter Bio-Dynamic) oversees the certification of biodynamic farms and products. Australian biodynamic farms produce a diverse array of crops, including grains, fruits, vegetables, and livestock. The country also hosts educational programs and conferences to promote biodynamic practices.

iv. **Europe:** Europe remains a significant hub for biodynamic farming. Countries such as France, Italy, Switzerland, and Austria have a strong tradition of biodynamic agriculture. In France, biodynamic viticulture has become particularly prominent, with many prestigious wineries adopting biodynamic methods to produce high-quality wines. The European Demeter Association coordinates biodynamic activities across the continent, ensuring adherence to standards and promoting research and education.

v. **North America:** In North America, biodynamic farming has found a niche within the broader organic farming community. The Biodynamic Association in the United States and Canada supports farmers through workshops, conferences, and certification programs. Biodynamic products, including wines, dairy, and produce, are increasingly available in specialty markets and through direct-to-consumer sales.

vi. **South America:** South America has witnessed growing interest in biodynamic farming, particularly in countries like Brazil, Argentina, and Chile. The region's fertile lands and favorable climate conditions provide an ideal environment for biodynamic agriculture. Several biodynamic farms in South America produce a wide range of crops, including coffee, wine, and tropical fruits, catering to both local and international markets.

vii. **Asia:** In Asia, biodynamic farming is gaining momentum, with countries like India, China, and Japan showing increasing interest. The biodynamic movement in Asia focuses on addressing the challenges of conventional agriculture, such as soil degradation, pesticide use, and loss of biodiversity. Biodynamic practices are seen as a way to restore ecological balance and improve food quality. Educational programs, farmer networks, and certification initiatives are being established to support the growth of biodynamic farming in the region.

3. Challenges and Future Prospects

Biodynamic farming, despite its numerous benefits and holistic approach, faces several challenges that impact its broader acceptance and implementation. However, the movement also holds promising future prospects that align with global trends towards sustainable agriculture and environmental conservation. Here, we delve into the challenges faced by biodynamic farming and explore its future prospects in detail.

4.1 Challenges

4.1.1 Skepticism and Lack of Scientific Validation

Biodynamic farming often encounters skepticism from the scientific community and conventional farmers. The spiritual and mystical aspects of biodynamic practices, such as the use of cosmic rhythms and biodynamic preparations, are sometimes viewed with suspicion and lack mainstream scientific validation. This skepticism can hinder the acceptance and integration of biodynamic methods within the broader agricultural and scientific communities.

1. **Labor-Intensive Practices:** Biodynamic farming is labor-intensive, requiring significant time and effort to implement and maintain its practices. Tasks such as preparing and applying biodynamic preparations, managing diverse crops and livestock, and adhering to lunar planting cycles demand meticulous attention and manual labor. Small-scale farmers with limited resources may find it challenging to adopt and sustain these labor-intensive practices.
2. **Access to Training and Education:** Access to training and education is crucial for farmers transitioning to biodynamic methods. However, in many regions, there is a lack of educational programs, workshops, and resources dedicated to biodynamic farming. Farmers may struggle to find reliable information and support to guide them in implementing biodynamic principles effectively.
3. **Economic Viability:** The economic viability of biodynamic farming can be a concern, especially for small-scale farmers. The initial costs of transitioning to biodynamic methods, such as acquiring certification, purchasing necessary inputs, and investing in infrastructure, can be substantial. Additionally, the potential for lower yields during the transition period may impact farmers' income, making it challenging to sustain their operations financially.
4. **Market Access and Consumer Awareness:** While there is a growing demand for sustainably produced food, biodynamic products often face challenges in market access and consumer awareness. Biodynamic

certification and labeling are less recognized compared to organic certification, leading to limited consumer understanding and demand for biodynamic products. This lack of market differentiation can impact farmers' ability to achieve premium prices for their biodynamic produce.

4.2 Future Prospects

1. **Growing Demand for Sustainable Agriculture:** The global trend towards sustainable agriculture presents a significant opportunity for biodynamic farming. Consumers are increasingly concerned about the environmental and health impacts of conventional farming practices and are seeking alternatives that prioritize sustainability, biodiversity, and soil health. Biodynamic farming, with its holistic and regenerative approach, aligns well with these consumer preferences, offering a promising future for the movement.
2. **Integration with Regenerative Agriculture:** Biodynamic farming's principles and practices resonate with the broader movement of regenerative agriculture. Regenerative agriculture focuses on restoring soil health, increasing biodiversity, and sequestering carbon to mitigate climate change. The alignment between biodynamic and regenerative agriculture provides an opportunity for collaboration and mutual reinforcement, enhancing the visibility and acceptance of biodynamic methods within the sustainable agriculture community.
3. **Advancements in Research and Validation:** Continued research and scientific validation of biodynamic practices are essential for addressing skepticism and gaining broader acceptance. Increasingly, studies are being conducted to investigate the efficacy of biodynamic preparations, the benefits of crop rotation and diversification, and the impact of lunar and cosmic rhythms on plant growth. Collaborative research efforts between biodynamic practitioners, scientists, and institutions can provide valuable insights and evidence to support the effectiveness of biodynamic farming.
4. **Educational Initiatives and Farmer Networks:** Expanding educational initiatives and farmer networks can play a crucial role in promoting biodynamic farming. Organizations such as the Biodynamic Association and Demeter can enhance their outreach by offering more workshops, training programs, and online resources. Establishing local farmer networks and mentorship programs can facilitate knowledge sharing, support new practitioners, and strengthen the biodynamic farming community.

5. **Market Development and Certification:** Efforts to increase market access and consumer awareness of biodynamic products are vital for the economic viability of biodynamic farming. Enhancing the recognition and credibility of biodynamic certification through marketing campaigns, partnerships with retailers, and participation in farmers' markets can help differentiate biodynamic products in the marketplace. Educating consumers about the benefits of biodynamic farming and the unique qualities of biodynamic produce can drive demand and support premium pricing.
6. **Policy Support and Incentives:** Government policies and incentives that support sustainable and regenerative agriculture can significantly benefit biodynamic farming. Policies that provide financial support, grants, and subsidies for transitioning to biodynamic methods, investing in infrastructure, and conducting research can alleviate some of the economic challenges faced by farmers. Additionally, policies that promote sustainable land use, conservation practices, and organic certification can create a more favorable environment for the growth of biodynamic farming.
7. **Technological Innovations:** Technological innovations can enhance the efficiency and effectiveness of biodynamic farming practices. For example, advancements in precision agriculture, remote sensing, and data analytics can help farmers monitor soil health, optimize crop rotations, and align planting schedules with lunar cycles more accurately. Integrating modern technologies with traditional biodynamic practices can improve productivity, reduce labor intensity, and enhance overall farm management.
8. **Global Collaboration and Knowledge Exchange:** International collaboration and knowledge exchange among biodynamic practitioners, researchers, and organizations can strengthen the global biodynamic movement. Sharing best practices, research findings, and innovations across borders can inspire new approaches and solutions to common challenges. Collaborative efforts can also help standardize biodynamic certification, promote global trade of biodynamic products, and foster a sense of community and shared purpose among biodynamic farmers worldwide.

Conclusion

Biodynamic farming, with its rich history and holistic approach, represents a transformative vision for agriculture. Rooted in the ideas of Rudolf Steiner, biodynamic practices have evolved into a globally recognized system that integrates ecological, spiritual, and cosmic principles. From its early development in Europe to its current international status, biodynamic farming continues to inspire farmers and consumers seeking sustainable and regenerative agricultural solutions.

The future of biodynamic farming lies in its ability to address the pressing challenges of modern agriculture, including soil degradation, biodiversity loss, and climate change. By fostering a deeper connection between humans and the natural world, biodynamic farming offers a pathway to a more sustainable and harmonious food system. Through continued research, education, and collaboration, the principles of biodynamic farming can contribute to the global movement towards a resilient and regenerative agricultural future.

References

Phillips, J. C., & Rodgriguez, L. P. (2006). Beyond organic: an overview of biodynamic agriculture with case examples.

Rehber, E. (2018). Organic farming: a historical perspective.

Schmid, O. (2007). Development of standards for organic farming. In Organic farming: An international history (pp. 152-174). Wallingford UK: CABI.

V+de Abreu, L. S., Bellon, S., & Corrales, F. M. (2005). The advent of organic farming models: analysis of the current situation and perspectives in Brazil. In: IFOAM ORGANIC WORLD CONGRESS, 15., 2005, Adelaide, Austrália. Proceedings... Bonn: ISOFAR, 2005. Editores Ulrich Köpke, Urs Niggli, Daniel Neuhoff. p. 611-613..

Wynen, E. (2007). Standards and compliance systems for organic and bio-dynamic agriculture in Australia: past, present and future. Journal of Organic Systems, 2(2), 38-55.

2

Principles, Concept and Comparison of Biodynamic Farming with other Farming Methods

Abstract

Biodynamic farming, an advanced form of organic agriculture, encompasses principles and practices aimed at fostering a holistic and self-sustaining ecosystem. Developed by Rudolf Steiner in the early 20th century, it integrates spiritual and cosmic elements with traditional farming methods to enhance soil health, plant vitality, and biodiversity. This article explores the core principles of biodynamic farming, including the farm-as-organism concept, biodynamic preparations, and the use of lunar and cosmic rhythms. It also provides a comparative analysis of biodynamic farming with other farming methods such as conventional, organic, and regenerative agriculture. The comparison highlights differences in sustainability, ecological impact, soil health, and crop quality. Through a comprehensive review, this article aims to elucidate the unique aspects of biodynamic farming, its benefits, challenges, and future prospects in the context of global agricultural practices.

Introduction

Biodynamic farming represents a unique and comprehensive approach to agriculture that integrates scientific and spiritual dimensions to cultivate not just crops, but entire ecosystems. Emerging in the early 20th century as a response to the growing concerns over the industrialization of agriculture, this method is grounded in the principles of anthroposophy, a philosophical movement founded by Rudolf Steiner. Biodynamic farming is not merely a set of agricultural practices; it embodies a worldview that considers the farm as a living, self-sustaining organism, where all elements are interrelated and interdependent.

At its core, biodynamic farming seeks to create a harmonious balance between the physical, biological, and spiritual aspects of farming. It emphasizes the importance of the farm's natural rhythms, including lunar and cosmic influences, which are believed to impact the growth and health of plants and animals.

DOI: 10.1201/9781042015597-2

Farmers who practice biodynamics use a variety of unique preparations made from fermented herbs, minerals, and manure to enhance soil fertility and plant vitality, following a biodynamic calendar that guides the timing of sowing, planting, and harvesting.

The principles of biodynamic farming extend beyond organic practices, which focus primarily on avoiding synthetic chemicals and genetically modified organisms. Biodynamics incorporates a more holistic perspective, advocating for closed-loop systems where all inputs, such as compost, seeds, and animal feed, are generated within the farm itself. This reduces the farm's dependency on external resources and promotes sustainability and resilience.

In comparison to other farming methods, biodynamic farming stands out for its spiritual and ecological focus. While organic farming shares some similarities, such as the avoidance of synthetic inputs, it does not typically involve the same level of attention to cosmic rhythms or the use of specialized preparations. Conventional farming, on the other hand, often relies on intensive inputs and practices that can lead to environmental degradation, making biodynamics a more sustainable alternative.

As the global agricultural landscape evolves in response to challenges like climate change, soil depletion, and food security, biodynamic farming offers a model that integrates environmental stewardship with a deeper connection to the natural world. This section will delve into the specific principles and concepts that define biodynamic farming, explore how it compares with other agricultural methods, and examine its potential to shape the future of sustainable farming.

1. Principles of Biodynamic Farming

1.1 Farm-as-Organism Concept

Biodynamic farming operates on the principle that the farm should be viewed as a self-contained and self-sustaining organism. This holistic perspective considers all components of the farm—soil, plants, animals, and humans—as interconnected parts of a single system. Each element supports and nourishes the others, creating a balanced and resilient ecosystem. This approach encourages farmers to foster biodiversity, enhance soil health, and integrate livestock to create a closed-loop system.

1.2 Biodynamic Preparations

One of the distinguishing features of biodynamic farming is the use of biodynamic preparations. These preparations, developed by Rudolf Steiner, are made from specific plant, animal, and mineral substances. They are used in small quantities to enhance soil fertility, stimulate plant growth, and improve compost quality. **Preparation 500 (Horn Manure)**: Made from cow manure buried in a cow horn over the winter, it is sprayed on the soil to enhance microbial activity and root development.

1.3 Lunar and Cosmic Rhythms

Biodynamic farming places significant emphasis on the influence of lunar and cosmic rhythms on plant growth and development. Farmers use a biodynamic planting calendar, which provides guidance on the best times to perform various agricultural activities, such as planting, cultivating, and harvesting. The calendar is based on the position of the moon, planets, and stars, with the belief that these celestial forces can affect the vitality of plants and soil.

2. Concept of Biodynamic Farming

Biodynamic farming is built on a set of core concepts that differentiate it from other forms of agriculture. These concepts include:

1. **Holistic View of the Farm**: The farm is seen as a living, self-sustaining organism where all elements are interconnected and contribute to the overall health and productivity of the system.

Fig. Illustration of Biodynamic Farming Concept

2. **Use of Biodynamic Preparations**: Biodynamic preparations are unique formulations made from natural substances that enhance soil fertility, plant growth, and compost quality.
3. **Alignment with Natural Cycles**: Biodynamic farming aligns agricultural activities with lunar and cosmic rhythms to optimize plant growth and vitality.
4. **Emphasis on Biodiversity**: Biodynamic farms typically grow a diverse range of crops and raise different types of animals, creating a balanced and resilient ecosystem.
5. **Focus on Soil Health**: Soil health is a central focus in biodynamic farming, with practices such as composting, crop rotation, and green manures used to enhance soil fertility and structure.

3. Comparison with Other Farming Methods

3.1 Conventional Farming

Conventional farming is characterized by the extensive use of synthetic inputs such as chemical fertilizers, pesticides, and herbicides to maximize crop yields and ensure pest control. While this method has played a significant role in meeting the global food demand, it often comes at the expense of environmental health. Prolonged reliance on synthetic chemicals leads to soil degradation, depletion of essential nutrients, loss of beneficial microorganisms, and contamination of water bodies through runoff. Additionally, monoculture practices, which are common in conventional farming, reduce biodiversity and make ecosystems more vulnerable to pests and diseases. In contrast, biodynamic farming avoids synthetic chemicals, relying instead on organic composts, manure, and plant-based preparations to nurture the soil. This holistic approach not only restores soil vitality but also fosters ecological harmony, making it a sustainable alternative to conventional farming.

3.2 Organic Farming

Organic farming and biodynamic farming share a foundational philosophy of sustainability and ecological stewardship. Both reject the use of synthetic inputs and emphasize the importance of maintaining soil health through natural methods, such as crop rotation, green manures, and composting. However, biodynamic farming distinguishes itself by integrating spiritual and cosmic principles into its practices. For instance, biodynamic preparations, such as fermented herbal and mineral mixtures, are used to enhance soil and plant vitality. These preparations are often applied following specific lunar and celestial cycles, which are believed to influence agricultural outcomes. Furthermore, biodynamic certification, governed by organizations like Demeter International, often imposes stricter standards than organic certification, focusing on whole-farm sustainability, biodiversity enhancement, and closed-loop nutrient cycles. This broader vision of interconnectedness sets biodynamic farming apart from organic methods.

3.3 Regenerative Agriculture

Regenerative agriculture represents a progressive approach to farming that seeks to rebuild and enhance the overall health of agricultural ecosystems. Its practices include cover cropping, no-till farming, agroforestry, and managed grazing, all aimed at improving soil health, sequestering carbon, and boosting biodiversity. Biodynamic farming aligns with many of these regenerative principles, particularly in its commitment to

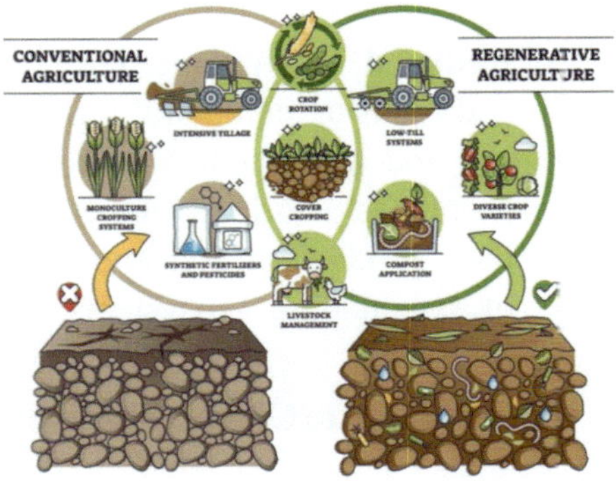

Fig. Comparision of conventional and regenerative farming

soil health and ecological balance. Both approaches recognize the farm as a living organism that thrives on diversity and interdependence. However, biodynamic farming introduces a unique spiritual dimension, incorporating cosmic rhythms and esoteric practices into its framework. For example, the timing of planting, harvesting, and application of biodynamic preparations is often guided by lunar and astrological calendars. This integration of spiritual awareness provides biodynamic farming with a distinct identity within the regenerative movement, offering an additional layer of holistic stewardship to agricultural practices.

4. Sustainability and Ecological Impact

Biodynamic farming's holistic approach to agriculture has significant implications for sustainability and ecological impact. By fostering biodiversity, enhancing soil health, and reducing reliance on synthetic inputs, biodynamic farming contributes to the creation of resilient and sustainable agricultural systems. Key benefits include:

1. **Improved Soil Health:** Biodynamic practices such as composting, crop rotation, and the use of biodynamic preparations enhance soil structure, fertility, and microbial activity, leading to healthier and more productive soils. This not only improves agricultural productivity but also reduces soil erosion and nutrient loss, creating a stable foundation for long-term farming success.
2. **Enhanced Biodiversity:** Biodynamic farms typically feature diverse crop rotations, mixed-species plantings, and integrated livestock systems. This diversity supports pollinators, natural pest predators, and a wide range of microorganisms, promoting ecological balance. By creating habitats for various species, biodynamic farming contributes to reversing the trend of biodiversity loss associated with conventional agricultural practices.
3. **Reduced Environmental Impact:** By avoiding synthetic chemicals and focusing on natural inputs, biodynamic farming reduces the risk of soil and water contamination. The absence of chemical fertilizers and pesticides also helps lower greenhouse gas emissions associated with their production and application. This approach supports the health of surrounding ecosystems, including rivers, lakes, and groundwater sources.
4. **Resilience to Climate Change:** The emphasis on soil health, biodiversity, and ecological balance in biodynamic farming helps create resilient agricultural systems that can better withstand the impacts of climate

change. Healthy soils with high organic matter content act as carbon sinks, helping to mitigate global warming. Furthermore, diversified farming systems are better equipped to adapt to extreme weather events, ensuring food security in the face of a changing climate.

Conclusion

Biodynamic farming is a holistic and regenerative approach to agriculture that integrates spiritual and cosmic principles with traditional farming methods. It offers numerous benefits, including improved soil health, enhanced biodiversity, and reduced environmental impact. However, it also faces challenges such as skepticism, labor intensity, and economic viability. The future prospects for biodynamic farming are promising, driven by the growing demand for sustainable agriculture, integration with regenerative agriculture, and advancements in research and validation. By addressing these challenges and leveraging future opportunities, biodynamic farming can contribute to a more sustainable, resilient, and harmonious agricultural system.

References

BELUHOVA-UZUNOVA, R.O.S.I.T.S.A., & ATANASOV, D. (2017). Biodynamic Farming-Method for Sustainable Production of Quality Food. Bulgarian Journal of Agricultural Economics & Management Ikonomikai Upravleniena Selskoto Stopanstvo, 62(3).

Kjellenberg, L., & Granstedt, A. (2015). Influences of biodynamic and conventional farming systems on quality of potato (Solanum tuberosum L.) crops: Results from multivariate analyses of two long-term field trials in Sweden. Foods, 4(3), 440

Muhie, S. H. (2023). Concepts, principles, and application of biodynamic farming: a review. Circular Economy and Sustainability, 3(1), 291-304.

Rigolot, C., & Quantin, M. (2022). Biodynamic farming as a resource for sustainability transformations: Potential and challenges. Agricultural Systems, 200, 103424.

Turinek, M., Grobelnik-Mlakar, S., Bavec, M., & Bavec, F.R.A.N.C. (2009). Biodynamic agriculture research progress and priorities. Renewable agriculture and food systems, 24(2), 146-154.

3

Biodynamic Preparations and Application of Preparataries

Abstract

Biodynamic farming is a holistic agricultural system that emphasizes the interrelationship between soil, plants, animals, and cosmic forces. A cornerstone of this practice is the use of biodynamic preparations, which are specialized herbal and mineral formulations applied to the soil, compost, and plants to enhance their health and vitality. This article delves into the origins, development, and specific types of biodynamic preparations, including their ingredients, preparation methods, and application techniques. By exploring both the scientific and spiritual dimensions of these preparations, the article aims to provide a comprehensive understanding of their role in biodynamic farming. It also examines the empirical evidence supporting their efficacy, addresses common criticisms, and discusses future research directions.

Introduction

Biodynamic farming is a holistic approach to agriculture that combines organic farming practices with spiritual and cosmic principles to create a self-sustaining and ecologically balanced farming system. One of the core components of biodynamic farming is the use of biodynamic preparations, which are specific substances made from natural materials designed to enhance soil fertility, plant health, and overall farm vitality. This article delves into the intricacies of biodynamic preparations, their types, methods of preparation, and their applications in biodynamic farming.

Biodynamic preparations are central to biodynamic farming, functioning as natural enhancers that stimulate biological activity and harmonize the farm ecosystem. These preparations were developed by Rudolf Steiner, the founder of biodynamic farming, and are used in small quantities to bring about significant improvements in soil health and plant growth.

There are two main categories of biodynamic preparations: field sprays and compost preparations. Field sprays are applied directly to the soil or plants to enhance soil fertility and plant growth, while compost preparations are added to compost piles to improve the quality of the compost and accelerate the decomposition process.

DOI: 10.1201/9781042015597-3

1. Field Spray Preparations

1.1 Preparation 500: Horn Manure (Cow Horn Manure)

Ingredients and Preparation:

- **Ingredients:** Fresh cow manure from cows that have calved at least once.
- **Preparation Method:** The cow manure is packed into cow horns and buried in the soil during the autumn. The horns are left buried over the winter, during which the manure undergoes a fermentation process. In the spring, the horns are unearthed, and the fermented manure, now called horn manure, is removed.
- **Application:** Horn manure is mixed with water and stirred for an hour to create a vortex. This mixture is then sprayed onto the soil. The purpose of horn manure is to stimulate soil microbial activity, enhance root growth, and improve soil structure. It is typically applied in the late afternoon or evening.

Fig. Horn Manure Preparation and Horn manure

1.2 Preparation 501: Horn Silica (Cow Horn Silica)

Ingredients and Preparation:

- **Ingredients:** Finely ground quartz or silica.
- **Preparation Method:** The silica is packed into cow horns and buried in the soil during the spring. The horns are left buried over the summer to absorb the sun's energy. In the autumn, the horns are unearthed, and the silica, now called horn silica, is removed.

Fig. Horn Silica

- **Application:** Horn silica is mixed with water and stirred for an hour to create a vortex. This mixture is then sprayed onto plants. The purpose of horn silica is to enhance photosynthesis, improve plant structure, and increase plant resistance to diseases. It is typically applied in the early morning.

2. Compost Preparations

Compost preparations are added to compost piles to enhance the decomposition process and enrich the compost with beneficial microorganisms. There are six main compost preparations, each made from specific plant materials.

2.1 Preparation 502: Yarrow (Achillea millefolium)

Ingredients and Preparation

- **Ingredients:** Yarrow flowers.
- **Preparation Method:** The yarrow flowers are harvested and placed in a stag's bladder, which is then hung in the sun during the summer and buried in the soil during the winter. The bladder is unearthed in the spring, and the yarrow preparation is removed.
- **Application:** Yarrow preparation is added to compost piles to enhance the decomposition process and enrich the compost with sulfur and potassium, which are essential for plant growth.

Fig. *Achillea millefolium*

2.2 Preparation 503: Chamomile (Matricaria recutita)

Ingredients and Preparation

- **Ingredients:** Chamomile flowers.
- **Preparation Method:** The chamomile flowers are harvested and placed in the intestines of a cow, which are then buried in the soil during the autumn. The intestines are unearthed in the spring, and the chamomile preparation is removed.
- **Application:** Chamomile preparation is added to compost piles to stabilize nitrogen, promote the breakdown of organic matter, and enhance the compost's overall quality.

Fig. *Matricaria recutita*

2.3 Preparation 504: Stinging Nettle (Urtica dioica)

Ingredients and Preparation:

- **Ingredients:** Stinging nettle leaves.
- **Preparation Method:** The stinging nettle leaves are harvested and placed in a pit in the soil, where they are left to decompose for a year.
- **Application:** Stinging nettle preparation is added to compost piles to stimulate soil health, enhance microbial activity, and promote the formation of humus.

Fig. *Urtica dioica*

2.4 Preparation 505: Oak Bark (Quercus robur)

Ingredients and Preparation:

- **Ingredients:** Oak bark.
- **Preparation Method:** The oak bark is harvested and placed in the skull of a domestic animal, which is then buried in a marshy area over the winter. The skull is unearthed in the spring, and the oak bark preparation is removed.
- **Application:** Oak bark preparation is added to compost piles to provide calcium, regulate the composting process, and promote plant health and disease resistance.

Fig. *Quercus robur*

2.5 Preparation 506: Dandelion (Taraxacum officinale)

Ingredients and Preparation:

- **Ingredients:** Dandelion flowers.
- **Preparation Method:** The dandelion flowers are harvested and placed in the mesentery (a membrane that attaches the intestines to the abdominal wall) of a cow, which is then buried in the soil over the winter. The mesentery is unearthed in the spring, and the dandelion preparation is removed.

Fig. *Taraxacum officinale*

- **Application:** Dandelion preparation is added to compost piles to enhance the uptake of silica by plants, promote root growth, and improve plant vitality.

2.6 Preparation 507: Valerian (Valeriana officinalis)

Ingredients and Preparation:

- **Ingredients:** Valerian flowers.
- **Preparation Method:** The valerian flowers are harvested and made into an extract by soaking them in water.
- **Application:** Valerian preparation is added to compost piles to stimulate phosphorus activity, promote composting, and enhance the compost's overall quality.

Fig. *Valeriana officinalis*

Application of Biodynamic Preparations

The application of biodynamic preparations is a meticulous process that involves precise timing, careful preparation, and thoughtful integration into the farm ecosystem. Below, we outline the steps for applying these preparations effectively.

3. Field Sprays

1. Preparation 500 (Horn Manure):

- **Timing:** Applied in the late afternoon or evening, typically in the autumn and spring.
- **Mixing:** Mix a small amount (about 1/4 to 1/2 cup) of horn manure with 13 to 30 gallons of water. Stir the mixture vigorously for an hour, creating a vortex by stirring in one direction and then reversing.
- **Application:** Using a sprayer or a brush, apply the mixture to the soil. Focus on areas where crops will be planted to enhance soil fertility and root development.

2. Preparation 501 (Horn Silica):

- **Timing:** Applied in the early morning, typically during the growing season.
- **Mixing:** Mix a small amount (about 1/4 to 1/2 teaspoon) of horn silica with 13 to 30 gallons of water. Stir the mixture vigorously for an hour, creating a vortex by stirring in one direction and then reversing.

- **Application:** Using a sprayer, apply the mixture to the foliage of plants. Ensure even coverage to enhance photosynthesis and plant vitality.

4. Compost Preparations

1. Adding to Compost Piles

- **Placement:** Create several small holes in the compost pile.
- **Addition:** Add small amounts of each compost preparation (502-507) into the holes. Cover the holes with compost to ensure the preparations are integrated into the pile.
- **Distribution:** Distribute the preparations evenly throughout the compost pile to promote balanced decomposition and nutrient enrichment.

2. Activating the Compost

- **Moisture Management:** Ensure the compost pile remains moist but not waterlogged. Moisture is crucial for microbial activity and the breakdown of organic matter.
- **Turning the Compost:** Regularly turn the compost pile to aerate it and promote even decomposition. This helps distribute the biodynamic preparations throughout the pile.
- **Monitoring:** Monitor the temperature and smell of the compost pile. A healthy compost pile should be warm and have an earthy smell. Adjust moisture and aeration as needed to maintain optimal conditions.

5. Benefits of Biodynamic Preparations

The use of biodynamic preparations offers numerous benefits for soil health, plant growth, and overall farm vitality. These benefits include:

1. **Enhanced Soil Fertility:** Biodynamic preparations stimulate soil microbial activity, enhance nutrient availability, and improve soil structure. This leads to healthier, more fertile soils that support robust plant growth.
2. **Improved Plant Health:** Biodynamic preparations enhance plant vitality, resilience, and resistance to diseases. Plants grown with biodynamic preparations often exhibit improved growth, higher yields, and better overall health.
3. **Balanced Ecosystem:** Biodynamic preparations contribute to the creation of a balanced and harmonious farm ecosystem. By promoting biodiversity, enhancing soil health, and aligning with natural cycles, biodynamic farming supports long-term ecological sustainability.

4. **Reduced Need for Synthetic Inputs:** The use of biodynamic preparations reduces the reliance on synthetic fertilizers, pesticides, and herbicides. This leads to lower input costs, reduced environmental impact, and healthier food products.
5. **Enhanced Compost Quality:** Compost prepared with biodynamic preparations decomposes more efficiently and results in high-quality compost rich in beneficial microorganisms and nutrients. This compost enhances soil fertility and plant health when applied to fields.

6. Challenges and Considerations

While biodynamic farming and the use of biodynamic preparations offer numerous benefits, they also present various challenges and considerations. Understanding and addressing these challenges is crucial for the successful implementation and sustainability of biodynamic practices. Below, we explore some of the primary challenges and considerations faced by biodynamic farmers.

6.1 Labor-Intensive Process

Preparation and Application: The preparation and application of biodynamic preparations are labor-intensive processes that require careful attention to detail and timing. For example, preparing field sprays like horn manure (Preparation 500) involves collecting fresh cow manure, packing it into cow horns, burying it in the soil for a specific period, and then retrieving and applying it. Similarly, compost preparations require meticulous preparation and integration into compost piles.

Ongoing Commitment: Biodynamic farming is not a one-time effort but an ongoing commitment. Farmers need to continuously prepare and apply these preparations according to specific lunar and seasonal cycles, which can be demanding and time-consuming.

Skill and Knowledge Requirements: Effective use of biodynamic preparations requires a deep understanding of the principles and practices of biodynamic farming. Farmers must be well-versed in the methods of preparation, timing of application, and monitoring of their effects. This necessitates ongoing education and training.

6.2 Access to Materials

Specific Materials Required: Some biodynamic preparations require specific materials that may not be readily available to all farmers. For instance, preparations like horn manure and horn silica require cow horns, while certain compost preparations need animal organs such as stag bladders, cow intestines, and skulls. Accessing these materials can be challenging, particularly for urban farmers or those in regions with limited livestock.

Quality of Ingredients: The quality of the ingredients used in biodynamic preparations is critical to their effectiveness. Farmers must source high-quality materials, which can be difficult and sometimes expensive. Ensuring the purity and integrity of these materials is essential for the desired outcomes.

Sourcing and Sustainability: In some cases, sourcing the required materials sustainably can be a concern. Farmers need to balance the demand for specific ingredients with ethical and sustainable sourcing practices to avoid environmental degradation or depletion of resources.

6.3 Skepticism and Acceptance

Scientific Validation: Biodynamic farming, with its spiritual and cosmic elements, often faces skepticism from the scientific community and conventional farmers. The principles and effectiveness of biodynamic preparations are not always supported by mainstream scientific research, leading to doubts about their legitimacy and efficacy.

Cultural Perceptions: The holistic and spiritual aspects of biodynamic farming may be perceived as unconventional or unscientific by some farmers and consumers. This can create barriers to acceptance and adoption, especially in regions with strong cultural preferences for conventional or organic farming methods.

Educational Outreach: Bridging the gap between biodynamic farming and conventional agriculture requires extensive educational outreach. Farmers and consumers need to be informed about the principles, benefits, and evidence supporting biodynamic practices. Demonstrating the tangible benefits through case studies, field trials, and research can help build credibility and acceptance.

6.4 Climate and Regional Adaptations

Variability in Effectiveness: The effectiveness of biodynamic preparations can vary based on climate, soil type, and regional conditions. What works well in one geographic area may not yield the same results in another. Farmers must adapt biodynamic practices to suit their specific environmental and climatic conditions.

Environmental Sensitivity: Biodynamic farming emphasizes working in harmony with natural cycles and environmental conditions. This requires farmers to be highly attuned to their local environment, observing and responding to changes in weather patterns, soil health, and plant growth. This sensitivity can be challenging, especially in regions with unpredictable or extreme weather conditions.

Customization and Flexibility: Biodynamic farmers need to be flexible and willing to experiment with different preparations and application methods to find what works best for their unique circumstances. This trial-and-error approach can be time-consuming and may require additional resources.

6.5 Economic Viability

Initial Investment: Transitioning to biodynamic farming can require significant initial investment in terms of time, labor, and resources. Farmers may need to purchase specific materials, invest in training and education, and potentially incur higher costs for biodynamic certification.

Market Challenges: While there is a growing demand for organic and sustainably produced food, biodynamic products still occupy a niche market. Farmers may face challenges in finding buyers willing to pay a premium for biodynamic produce. Building consumer awareness and demand for biodynamic products is essential for economic sustainability.

Certification Costs: Obtaining biodynamic certification through organizations like Demeter can involve additional costs and administrative burdens. Farmers must adhere to strict standards and undergo regular inspections, which can be financially and logistically challenging.

6.6 Regulatory and Policy Support

Lack of Government Support: In many regions, biodynamic farming does not receive the same level of government support or subsidies as conventional agriculture. This lack of policy support can make it harder for biodynamic farmers to compete economically.

Policy Advocacy: Advocating for policies that support biodynamic and other regenerative farming practices is crucial. Farmers and organizations need to work together to influence agricultural policies, securing incentives, grants, and subsidies that promote sustainable farming methods.

6.7 Community and Network Building

Isolation of Practitioners: Biodynamic farmers may often feel isolated due to the niche nature of their practice. Building strong networks and communities of biodynamic practitioners can provide essential support, knowledge exchange, and encouragement.

Sharing Best Practices: Collaborative efforts to share best practices, research findings, and success stories can help build a stronger and more resilient biodynamic farming community. Farmer-to-farmer learning, mentorship programs, and regional biodynamic associations can play a vital role in this regard.

Conclusion

Biodynamic preparations are a cornerstone of biodynamic farming, offering a holistic approach to enhancing soil fertility, plant health, and overall farm vitality. These preparations, derived from natural materials and crafted through specific processes, embody the principles of biodynamic agriculture by promoting ecological balance, sustainability, and harmony with natural cycles. Through careful preparation and thoughtful application, biodynamic preparations can transform farming practices, leading to healthier soils, more resilient plants, and a more sustainable agricultural system. While the use of these preparations presents certain challenges, their benefits make them a valuable tool for farmers committed to regenerative and sustainable farming practices.

As the global demand for sustainable and ecologically responsible agriculture continues to grow, biodynamic preparations offer a promising pathway toward a more harmonious and resilient farming future. By embracing the principles of biodynamic farming and integrating these preparations into their practices, farmers can contribute to the creation of a healthier, more sustainable planet.

References

Carpenter-Boggs, L., Kennedy, A. C., &Reganold, J. P. (2000). Organic and biodynamic management effects on soil biology. Soil Science Society of America Journal, 64(5), 1651-1659.

Carpenter-Boggs, L., Reganold, J. P., & Kennedy, A. C. (2000). Biodynamic preparations: Short-term effects on crops, soils, and weed populations. American Journal of Alternative Agriculture, 15(3), 110-118.

Paoletti, F., N. Nardo, I. Baiamonte, M. S. Foddai, E. Azzini, E. Finotti, A. Aguzzi, L. Gambelli, S. Di Ferdinando, and S. Paoletti. "Effects of biodynamic preparations on quality characteristics of organically grown tomato." Journal of Nutritional Ecology and Food Research 3, no. 2 (2016): 161-167.

Sedlmayr, A., van Leewen, A., Schönfelder, J., Kolar, M., Ingold, R., & Hurter, U. (2016). The biodynamic preparations in context: Individual approaches to preparation work-Case studies of worldwide practice. Goetheanum, Sektion für Landwirtschaft.

Spaccini, R., Mazzei, P., Squartini, A., Giannattasio, M., & Piccolo, A. (2012). Molecular properties of a fermented manure preparation used as field spray in biodynamic agriculture. Environmental Science and Pollution Research, 19, 4214-4225.

Vlahova, V., &Arabska, E. (2015). Biodynamic preparations–an alternative in the sustainable agricultural system. New knowledge Journal of science, 4(3).

4

Soil Health and Fertility in Biodynamic Farming

Abstract

Soil health and fertility are fundamental to sustainable agriculture, and biodynamic farming offers a holistic approach to enhancing these aspects. Biodynamic farming, developed by Rudolf Steiner in the early 20th century, integrates organic principles with cosmic and spiritual elements to create a self-sustaining and ecologically balanced farming system. This article explores the concept of soil health and fertility in the context of biodynamic farming, emphasizing the principles and practices that contribute to soil vitality. It discusses the importance of soil health, the key biodynamic practices for enhancing soil fertility, and the benefits and challenges associated with these practices. By integrating biodynamic preparations, composting techniques, and crop management strategies, biodynamic farming aims to restore and maintain soil health, leading to sustainable agricultural practices and improved farm ecosystems.

Introduction

Soil health and fertility are pivotal to the success and sustainability of agricultural systems. The condition of soil directly influences plant growth, crop yield, and overall farm productivity. A focus on soil health encompasses not just the physical and chemical properties of soil but also its biological vitality, which includes the diverse community of microorganisms, fungi, and fauna that interact within the soil ecosystem. Biodynamic farming, founded by Rudolf Steiner in the early 20th century, offers a distinctive approach to soil health and fertility, integrating organic principles with a holistic view of agriculture that encompasses cosmic and spiritual dimensions.

Biodynamic farming is an agricultural philosophy that extends beyond conventional organic practices. It views the farm as a living organism, where soil, plants, animals, and cosmic forces are interconnected and interdependent. This perspective emphasizes that the health of the soil is intrinsically linked to the broader ecological and cosmic environment. According to biodynamic principles, soil fertility is not merely a function of nutrient inputs but a reflection of the overall balance and vitality of the farm ecosystem.

DOI: 10.1201/9781042015597-4

At the heart of biodynamic farming are specific practices designed to enhance soil fertility and promote ecological balance. These include the use of biodynamic preparations, composting techniques, and crop management strategies that align with natural rhythms and cycles. Biodynamic preparations, made from natural materials such as plants, minerals, and animal substances, are applied to the soil and compost to stimulate biological activity and improve soil health. These preparations are believed to work in harmony with cosmic forces, enhancing the soil's ability to support plant growth and resilience.

Composting is another cornerstone of biodynamic soil management. Biodynamic composting involves the careful preparation and application of compost, often using biodynamic preparations to enrich the compost with beneficial microorganisms and nutrients. This process helps build soil structure, improve moisture retention, and increase nutrient availability. Composting also contributes to the recycling of organic matter, fostering a self-sustaining and nutrient-rich soil environment.

Crop rotation and diversity are essential practices in biodynamic farming for maintaining soil fertility. Rotating crops helps prevent the depletion of specific nutrients and reduces the risk of soil-borne diseases and pests. Introducing a variety of plant species, including cover crops and legumes, supports soil health by enhancing nutrient cycling, promoting beneficial interactions between plants and soil microorganisms, and preventing soil erosion.

Effective soil management in biodynamic farming involves regular monitoring of soil conditions, incorporating organic matter, and minimizing soil disturbance. Biodynamic farmers pay close attention to soil texture, structure, and microbial activity, adjusting their practices based on observations and soil tests. By incorporating organic matter such as compost and green manures, farmers enhance soil fertility and support healthy soil ecosystems. Additionally, reducing tillage and other forms of soil disturbance helps preserve soil structure and function.

The integration of these practices reflects the biodynamic commitment to creating a harmonious and sustainable farming system. While challenges such as labor intensity, access to materials, and scientific validation exist, the benefits of improved soil health, increased biodiversity, and reduced reliance on synthetic inputs underscore the value of biodynamic farming. By fostering a deeper understanding of soil health and fertility through biodynamic practices, farmers can contribute to a more resilient and ecologically balanced agricultural system, ultimately benefiting both the environment and food production.

1. Principles of Soil Health in Biodynamic Farming

1.1 Holistic Approach

Biodynamic farming views the farm as a self-sustaining organism where all components—soil, plants, animals, and cosmic influences—are interconnected. This holistic approach emphasizes the importance of balancing these elements to enhance soil health and fertility.

1.2 Ecological Balance

Maintaining ecological balance is central to biodynamic farming. Healthy soils are seen as a reflection of a balanced farm ecosystem. Biodynamic practices aim to support soil microorganisms, promote nutrient cycling, and foster a diverse soil fauna to achieve this balance.

1.3 Cosmic Influences

Biodynamic farming incorporates cosmic rhythms and cycles into its practices. Steiner proposed that lunar phases and planetary movements influence plant growth, soil processes, and overall farm health. By aligning agricultural activities with these cosmic rhythms, biodynamic farming seeks to optimize soil fertility and crop production.

2. Key Biodynamic Practices for Enhancing Soil Fertility

Biodynamic farming integrates a range of practices to enhance soil fertility, reflecting its holistic approach to agriculture. These practices are designed to promote soil health, foster ecological balance, and support sustainable farming systems. Here, we elaborate on the key biodynamic practices for enhancing soil fertility, focusing on biodynamic preparations, composting techniques, crop rotation, and soil management.

1. Biodynamic Preparations

Biodynamic preparations are integral to biodynamic farming. They are natural substances prepared from specific plant and animal materials and applied to the soil or compost to stimulate biological activity and enhance soil fertility.

Field Spray Preparations

- **Preparation 500 (Horn Manure):** This preparation involves packing fresh cow manure into cow horns and burying them in the soil for several months. The manure ferments inside the horns and is then diluted in water and sprayed onto the soil. Preparation 500 is believed to improve soil structure, enhance microbial activity, and stimulate root growth. The application of this preparation aims to balance soil moisture and increase the soil's ability to absorb and retain nutrients.

- **Preparation 501 (Horn Silica):** Made from finely ground quartz, Preparation 501 is packed into cow horns and buried in the soil. After several months, it is retrieved, diluted in water, and sprayed onto plants and soils. This preparation is thought to enhance photosynthesis, strengthen plant tissues, and improve plant resistance to diseases. It is used primarily to boost the plant's ability to utilize light and energy from the sun.

Compost Preparations

- **Preparation 502 (Yarrow):** Yarrow flowers are used in this preparation to help balance sulfur and potassium in the compost. It promotes the formation of humus and supports the decomposition process, enriching the compost with essential nutrients.
- **Preparation 503 (Chamomile):** Chamomile is added to compost to stabilize nitrogen and enhance the compost's microbial activity. This preparation aids in the breakdown of organic materials and improves the quality of the compost.
- **Preparation 504 (Stinging Nettle):** Nettle is used to activate microbial activity in compost and promote the formation of humus. It also helps in the development of plant root systems and enhances soil fertility.
- **Preparation 505 (Oak Bark):** Oak bark is used to regulate the composting process and provide calcium to the compost pile. It supports the development of soil structure and fertility.
- **Preparation 506 (Dandelion):** Dandelion preparation enhances silica uptake and promotes root growth. It is used to improve the resilience of plants and support their development.
- **Preparation 507 (Valerian):** Valerian preparation stimulates phosphorus activity in compost, which is vital for root development and plant growth. It also helps balance the compost's nutrient content.

2. Composting Techniques

Effective composting is a cornerstone of soil fertility in biodynamic farming. Composting enhances soil structure, adds organic matter, and supplies essential nutrients to the soil.

- **Material Selection:** Use a diverse mix of plant and animal materials, such as green and brown plant matter, animal manures, and other organic materials. This diversity helps ensure a balanced nutrient profile and promotes healthy microbial activity.

- **Preparation and Inoculation:** Incorporate biodynamic compost preparations into the compost pile. These preparations stimulate microbial activity and improve the compost's effectiveness. For example, adding Preparation 502 (Yarrow) and Preparation 503 (Chamomile) helps balance nutrients and enhance decomposition.
- **Pile Management:** Regularly turn the compost pile to ensure even decomposition and aeration. Monitor moisture levels and adjust as needed to maintain optimal conditions for microbial activity.
- **Application:** Once the compost is fully decomposed, apply it to the soil to improve its structure, fertility, and microbial content. Compost can be spread on fields, used as mulch, or incorporated into the soil.

3. Crop Rotation and Diversity

Crop rotation and diversity are essential practices in biodynamic farming to maintain soil fertility and prevent soil depletion.

3.1 Crop Rotation

- **Principle:** Rotating crops helps prevent the build-up of pests and diseases, reduces soil depletion, and supports nutrient cycling. Different plants have varying nutrient requirements and root structures, which contribute to a balanced soil profile.
- **Implementation:** Plan rotations based on plant families and their nutrient needs. For example, legumes fix nitrogen in the soil, while root crops help break up compacted soil layers. Rotating these crops can improve soil health and reduce the need for synthetic fertilizers.

3.2 Plant Diversity

- **Principle:** Planting a diverse range of crops, including cover crops and companion plants, supports soil health and promotes beneficial interactions between plants, soil microorganisms, and insects.
- **Implementation:** Incorporate a variety of plant species, such as legumes, grasses, and flowering plants, into the farming system. Cover crops, such as clover and vetch, can enhance soil fertility by adding organic matter and fixing nitrogen.

4. Soil Management

Effective soil management practices are critical for maintaining and enhancing soil fertility in biodynamic farming.

4.1 Soil Monitoring

- **Principle:** Regularly monitor soil health indicators, such as texture, structure, and nutrient levels, to assess soil conditions and make informed management decisions.
- **Implementation:** Use soil tests and visual observations to evaluate soil properties. Adjust soil management practices based on the results to address any deficiencies or imbalances.

4.2 Organic Matter Management

- **Principle:** Adding organic matter to the soil improves soil structure, moisture retention, and nutrient availability. Organic matter also supports beneficial soil microorganisms and enhances soil fertility.
- **Implementation:** Incorporate compost, green manures, and other organic amendments into the soil. Apply organic matter regularly to maintain soil health and support plant growth.

4.3 Minimizing Soil Disturbance

- **Principle:** Avoid excessive tillage and soil disturbance to protect soil structure and promote healthy soil ecosystems. Reduced soil disturbance helps maintain soil fertility and prevents erosion.
- **Implementation:** Use reduced tillage techniques, such as no-till or minimal tillage, to protect soil structure. Implement practices that reduce soil compaction and erosion, such as contour plowing and cover cropping.

5. Benefits of Biodynamic Soil Practices

1. **Improved Soil Structure:** Biodynamic practices enhance soil structure by promoting the formation of aggregates and improving soil aeration and drainage. This leads to better root growth and increased water-holding capacity.
2. **Enhanced Soil Fertility:** The use of biodynamic preparations and compost enriches the soil with essential nutrients and organic matter. This improves nutrient availability and supports healthy plant growth.
3. **Increased Soil Biodiversity:** Biodynamic farming promotes soil biodiversity by supporting a diverse range of soil microorganisms, fungi, and fauna. This enhances nutrient cycling, soil health, and ecosystem resilience.
4. **Reduced Soil Erosion:** Healthy soils with improved structure and organic matter content are less prone to erosion. Biodynamic practices help protect the soil from erosion and degradation.

5. **Sustainable Farming System:** Biodynamic farming creates a self-sustaining and ecologically balanced farming system. By focusing on soil health and fertility, it reduces the need for synthetic inputs and promotes long-term sustainability.

5. Challenges and Considerations

1. **Labor and Resource Intensity:** Biodynamic farming requires significant labor and resources for preparing and applying biodynamic preparations, composting, and managing soil health. This can be a challenge for farmers with limited time or resources.
2. **Access to Materials:** Obtaining specific materials for biodynamic preparations, such as cow horns or plant parts, can be challenging, particularly for farmers in urban or remote areas.
3. **Scientific Validation:** The effectiveness of biodynamic practices and preparations is not always supported by mainstream scientific research. This can lead to skepticism and challenges in gaining broader acceptance.
4. **Economic Viability:** The cost of implementing biodynamic practices, including certification and preparation materials, may be higher than conventional farming methods. Farmers need to assess the economic viability and potential return on investment.
5. **Climate and Regional Adaptations:** Biodynamic practices may need to be adapted to suit different climate and soil conditions. Farmers must tailor their approaches based on local environmental factors.

7. Future Directions

1. **Research and Development:** Ongoing research is needed to validate and refine biodynamic practices. Collaborative research efforts can provide evidence of their effectiveness and address scientific skepticism.
2. **Education and Training:** Expanding education and training opportunities for farmers can enhance the adoption and implementation of biodynamic practices. Workshops, field days, and online resources can support knowledge sharing and skill development.
3. **Policy Support:** Advocating for policies that support biodynamic and sustainable agriculture can provide financial incentives, grants, and subsidies to farmers. This can help overcome economic challenges and promote wider adoption.
4. **Technology Integration:** Integrating technology, such as precision agriculture tools and data analytics, can enhance the effectiveness of biodynamic practices. This includes monitoring soil conditions, optimizing nutrient applications, and improving farm management.

5. **Global Collaboration:** nternational collaboration and knowledge exchange among biodynamic practitioners can strengthen the global biodynamic movement. Sharing best practices, research findings, and success stories can inspire new approaches and solutions.

Conclusion

Soil health and fertility are crucial to the success and sustainability of biodynamic farming. By integrating holistic principles, biodynamic preparations, composting techniques, and effective soil management practices, biodynamic farming aims to restore and enhance soil vitality. While challenges such as labor intensity, material access, and scientific validation exist, the benefits of improved soil structure, fertility, biodiversity, and sustainability make biodynamic farming a valuable approach to agriculture. With continued research, education, and policy support, biodynamic practices can contribute to a more resilient and sustainable farming future.

References

Krause, H. M., Stehle, B., Mayer, J., Mayer, M., Steffens, M., Mäder, P., &Fliessbach, A. (2022). Biological soil quality and soil organic carbon change in biodynamic, organic, and conventional farming systems after 42 years. Agronomy for Sustainable Development, 42(6), 117.

Ponzio, C., Gangatharan, R., & Neri, D. (2013). Organic and biodynamic agriculture: a review in relation to sustainability. International Journal of Plant & Soil Science, 2(1), 95-110.

Reganold, J. P. (1995). Soil quality and profitability of biodynamic and conventional farming systems: A review. American journal of alternative agriculture, 10(1), 36-45.

Santoni, M., Ferretti, L., Migliorini, P., Vazzana, C., & Pacini, G. C. (2022). A review of scientific research on biodynamic agriculture. Organic Agriculture, 12(3), 373-396

5

Influence of Lunar and Astrological Rhythms Planting and Cultivation Practices

Abstract

The influence of lunar and astrological rhythms on agricultural practices has been a subject of interest and practice for centuries. In biodynamic farming, these celestial rhythms play a significant role in guiding planting and cultivation activities. Biodynamic farming, founded by Rudolf Steiner, integrates cosmic rhythms, including lunar phases and astrological cycles, into its agricultural practices to enhance soil fertility, plant growth, and overall farm health. This article explores the principles underlying the influence of lunar and astrological rhythms on planting and cultivation practices, the historical context of these practices, and their application in modern biodynamic farming. It examines how lunar and astrological cycles are believed to affect plant growth, soil health, and pest management, and provides insights into practical applications and benefits. By understanding the interplay between cosmic rhythms and agricultural practices, farmers can optimize their cultivation strategies to align with natural cycles, potentially leading to more sustainable and productive farming systems.

Introduction

The influence of lunar and astrological rhythms on planting and cultivation practices is one of the most distinctive and intriguing aspects of biodynamic farming. This approach, rooted in ancient agricultural traditions and refined by the teachings of Rudolf Steiner, reflects a profound belief in the interconnectedness of the cosmos and the natural world. While modern agriculture often emphasizes technology and science, biodynamic farming reintroduces the idea that celestial forces play a vital role in the growth and health of plants.

Central to this belief is the understanding that the moon, much like its well-documented influence on tides, also affects the water content in soil and plants. Biodynamic practitioners observe the moon's phases and its position relative to the constellations, using this information to time various agricultural

DOI: 10.1201/9781042015597-5

activities. For instance, specific periods in the lunar cycle are believed to be more favorable for sowing seeds, pruning, and harvesting, based on how the moon's gravitational pull impacts plant sap flow, root growth, and overall vitality.

Astrological rhythms further extend this concept, where the position of planets and their alignments are considered to influence the quality and characteristics of plant growth. Biodynamic calendars, which are meticulously developed each year, provide detailed guidance on when to perform different farming tasks, aligning these activities with the cosmic forces believed to be at play. This practice is not just about timing; it embodies a holistic approach that seeks to harmonize agricultural practices with the rhythms of the Earth and the wider universe. Critics may view the reliance on lunar and astrological rhythms as unscientific, but many biodynamic farmers report tangible benefits, such as improved crop yields, enhanced soil health, and increased plant resilience. The practice encourages farmers to be more attuned to the natural cycles around them, fostering a deeper connection with the land and a more intuitive approach to cultivation.

In this section, we will explore how lunar and astrological rhythms influence planting and cultivation practices in biodynamic farming. We will examine the science behind these beliefs, the specific techniques used by farmers, and how this ancient wisdom is being integrated into modern sustainable agriculture. This exploration will provide insight into a farming method that not only nourishes the soil and plants but also connects the farmer to the broader cosmic environment.

1. Historical Context and Theoretical Foundations

1.1 Historical Use of Celestial Rhythms in Agriculture

The use of lunar and astrological rhythms in agriculture can be traced back to ancient civilizations, including the Egyptians, Babylonians, and Mayans. These cultures observed the natural cycles of the moon and planets and integrated them into their agricultural practices. For example, the timing of planting and harvesting was often aligned with lunar phases, believed to influence soil and plant vitality.

1.2 Theoretical Foundations of Biodynamic Farming

Rudolf Steiner introduced biodynamic farming in the early 20th century, emphasizing the integration of cosmic rhythms into agricultural practices. Steiner's approach is based on the idea that the moon, planets, and stars have a direct influence on plant growth and soil health. According to biodynamic principles, aligning farming activities with these cosmic rhythms can enhance soil fertility, plant development, and overall farm resilience.

2. Lunar Rhythms and Their Influence on Plant Growth

2.1 Lunar Phases and Planting

Lunar phases—new moon, waxing moon, full moon, and waning moon—are believed to influence plant growth and development. Each phase is thought to affect different aspects of plant health:

- **New Moon:** Associated with the start of the lunar cycle, the new moon is considered an optimal time for planting root crops. The gravitational pull of the moon is thought to encourage downward growth, promoting strong root development.
- **Waxing Moon:** During the waxing moon, which occurs from the new moon to the full moon, plant growth is believed to be vigorous. This phase is considered favorable for planting above-ground crops, as the increased light and energy support leafy growth and flowering.
- **Full Moon:** The full moon is associated with peak growth and vitality. It is thought to be a good time for harvesting and weeding, as plants are believed to be at their most robust and resilient.
- **Waning Moon:** From the full moon to the new moon, the waning moon is considered a period for soil preparation and planting root crops. The decreasing light is thought to encourage inward growth and root development.

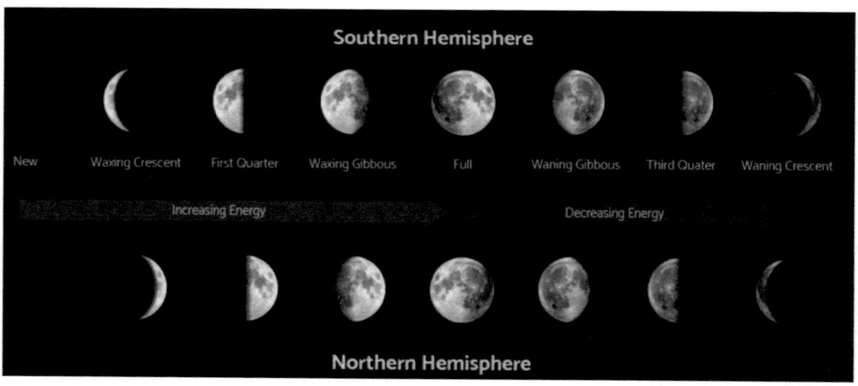

2.2 Lunar Influences on Soil Health

Lunar cycles are also believed to impact soil health and moisture levels. The gravitational pull of the moon is thought to influence soil moisture, with higher moisture levels during the full moon and lower levels during the new moon. This can affect soil preparation and irrigation practices, with potential implications for soil fertility and plant growth.

3. Astrological Cycles and Their Impact on Agriculture

3.1 Zodiac Signs and Planting

Biodynamic farming is a holistic agricultural approach that incorporates cosmic rhythms, lunar cycles, and astrological influences into cultivation methods. One key aspect of this practice is the use of zodiac signs to determine optimal planting, cultivation, and harvesting times. This method aligns farming activities with the energetic qualities of the zodiac, aiming to enhance plant vitality, soil health, and overall farm productivity. Here's an in-depth exploration of how zodiac signs influence planting in biodynamic farming:

1. **Earth Signs: Taurus, Virgo, Capricorn:** Earth signs in astrology are associated with stability, structure, and nourishment. In the context of biodynamic farming, these signs are ideal for tasks that promote soil fertility and root development. The grounded and practical energies of earth signs make them suitable for activities that involve physical connection to the soil.

Characteristics and Their Relevance

- **Taurus:** Known for its steady and nurturing qualities, Taurus is ideal for planting root crops such as potatoes, carrots, and beets. These crops benefit from the earth sign's focus on growth beneath the surface.
- **Virgo:** Virgo's precision and attention to detail align with activities like weeding, soil preparation, and planting herbs or vegetables that require meticulous care.
- **Capricorn:** As a sign associated with long-term planning and structure, Capricorn days are excellent for planting perennials, trees, and other crops that require stable growth over time.

2. Water Signs: Cancer, Scorpio, Pisces

Water signs represent fluidity, intuition, and emotional depth. In biodynamic farming, these signs are closely linked to moisture, which is crucial for leafy greens and crops requiring abundant water.

Characteristics and Their Relevance

- **Cancer:** With its nurturing and protective qualities, Cancer is ideal for planting leafy vegetables like lettuce, spinach, and cabbage. Its influence supports hydration and growth above ground.
- **Scorpio:** Known for its intensity and transformative energy, Scorpio days are suitable for planting crops that require deep rooting, such as tubers and vines, ensuring strong anchorage and nutrient uptake.

- **Pisces:** The intuitive and adaptable nature of Pisces aligns with planting water-intensive crops like rice or melons. It also supports the cultivation of plants used in medicinal and aromatic applications.

3. Air Signs: Gemini, Libra, Aquarius

Air signs symbolize movement, communication, and intellectual growth. In farming, their influence is beneficial for activities that promote flowering, pollination, and air circulation around plants.

Characteristics and Their Relevance

- **Gemini:** The dualistic and dynamic energy of Gemini supports the growth of flowering plants and herbs that thrive in light, airy conditions, such as chamomile and lavender.
- **Libra:** Libra's balance and harmony make it an ideal time for planting ornamental flowers, ensuring aesthetic appeal and structural symmetry in gardens.
- **Aquarius:** As a forward-thinking and innovative sign, Aquarius days are suitable for experimenting with new planting techniques or cultivating crops that benefit from aeration and innovation, like hydroponic plants.

4. Fire Signs: Aries, Leo, Sagittarius

Fire signs embody energy, enthusiasm, and vitality. Their influence in biodynamic farming is associated with planting crops that demand strong growth and thrive under warm conditions.

Characteristics and Their Relevance

- **Aries:** The pioneering and energetic qualities of Aries make it suitable for planting early-season crops that require a quick start, such as radishes and peas.
- **Leo:** Leo's vibrant and life-giving energy supports the growth of sun-loving crops, including tomatoes, peppers, and sunflowers. It is also a good time for harvesting fruits.
- **Sagittarius:** Known for its expansive and adventurous spirit, Sagittarius is ideal for planting climbing plants like beans and grapes, as well as crops intended for long-term storage.

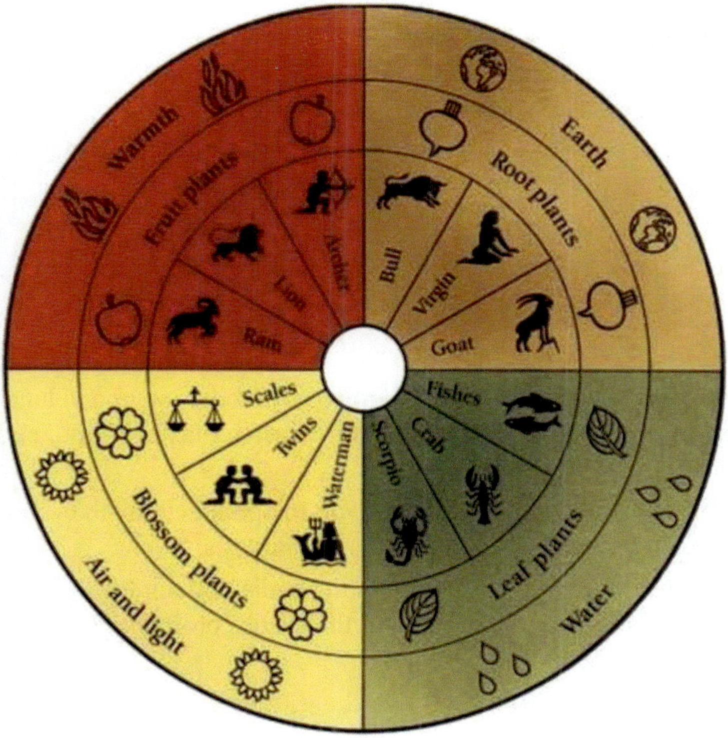

Fig. Illustration of Zodiac Signs and Agriculture planting

4. Integrating Zodiac Signs into Biodynamic Practices

Biodynamic farming emphasizes harmony between the celestial and terrestrial realms. Farmers following this approach use biodynamic calendars to synchronize agricultural activities with zodiac signs and lunar phases. Here are some practical steps to integrate zodiac influences into farming:

1. **Consult a Biodynamic Calendar:** Biodynamic calendars highlight the optimal days for sowing, transplanting, pruning, and harvesting based on the moon's position in the zodiac. These calendars act as guides to align farming activities with cosmic rhythms.
2. **Understand Plant Groups:** Different plant types respond to specific zodiac energies:
   - **Root Crops:** Favorable under earth signs.
   - **Leafy Greens:** Best planted under water signs.
   - **Flowering Plants:** Thrive under air signs.
   - **Fruit Crops:** Prosper under fire signs.

3. **Combine with Lunar Cycles:** The moon's waxing and waning phases further enhance the zodiac's influence. For instance:
   - **Waxing Moon:** Promotes upward growth, making it ideal for planting above-ground crops.
   - **Waning Moon:** Encourages root development and is suitable for underground crops.
4. **Monitor Soil and Weather Conditions:** While zodiac signs provide a cosmic guide, practical considerations like soil health, moisture levels, and weather patterns must be factored into decision-making.

4.1 Benefits of Zodiac-Based Planting

1. **Enhanced Crop Quality:** Aligning planting with zodiac signs and lunar phases can result in healthier plants with improved taste, aroma, and nutritional value.
2. **Increased Yield:** Biodynamic practices aim to optimize growth cycles, leading to better productivity and sustainable harvests.
3. **Holistic Farm Management:** Incorporating zodiac principles fosters a deeper connection between farmers and their land, promoting mindful cultivation.
4. **Environmental Sustainability:** By working in harmony with natural cycles, biodynamic farming reduces the need for synthetic inputs and supports ecological balance.

4.2 Challenges and Misconceptions

While biodynamic farming offers numerous benefits, it also faces challenges and skepticism:

1. **Scientific Validation:** The astrological basis of biodynamic farming lacks extensive scientific backing, leading some to dismiss it as pseudoscience.
2. **Complexity:** Following a biodynamic calendar requires careful planning and a thorough understanding of cosmic rhythms, which can be daunting for new practitioners.
3. **Regional Variations:** Climatic and geographic differences may influence the applicability of zodiac-based planting in certain areas.

5. Planetary Influences and Agricultural Timing

Biodynamic farming, rooted in holistic and ecological principles, emphasizes the interplay between cosmic forces and agricultural practices. Among these cosmic elements, planetary influences play a pivotal role in determining the optimal timing for various farming activities. The belief that planetary alignments can directly impact soil health, plant growth, and pest dynamics forms the foundation of this approach. This chapter explores how the movements of Venus, Mars, and Saturn shape key aspects of biodynamic farming.

1. **Venus: The Planet of Growth and Fertility:** In biodynamic farming, Venus is traditionally associated with fertility and growth. Its influence is considered most favorable during periods when the planet aligns harmoniously with the Earth, creating conditions that enhance plant vigor and soil vitality. The following are specific activities linked to Venus:
   - **Planting and Cultivation:** Venus is believed to govern the optimal times for sowing seeds, particularly those of crops that demand high levels of nourishment, such as leafy greens and vegetables. Farmers often align their planting schedules with Venusian cycles to ensure robust growth and abundant yields.
   - **Flowering Plants:** The planet's energy is also thought to favor the cultivation of flowers, both for ornamental purposes and for use in herbal remedies. By timing planting and pruning activities with Venus's influence, growers aim to maximize blooms and enhance the aromatic qualities of flowers.
   - **Soil Enrichment:** Biodynamic preparations such as compost teas and herbal sprays are often applied during Venusian periods. The fertility-promoting energies of Venus are believed to enhance the absorption and efficacy of these treatments, improving soil structure and nutrient availability.
2. **Mars: Energy and Action:** Mars, characterized by its association with energy, action, and intensity, is thought to influence agricultural practices requiring vigor and assertiveness. In biodynamic farming, Mars is often consulted for its role in managing pests, weeds, and diseases. Key activities under Mars's influence include:
   - **Pest Control:** Farmers align pest management efforts with Mars's cycles, as the planet's energy is believed to enhance the effectiveness of natural repellents and traps. Activities such as releasing beneficial insects or applying pest-deterrent sprays are often timed to coincide with Mars's active phases.
   - **Disease Management:** The fiery energy of Mars is also considered beneficial for combating plant diseases. Treatments targeting fungal

infections or bacterial outbreaks are believed to be more effective when administered during Mars's periods of prominence.

- **Weed Removal:** The assertive qualities of Mars make it an ideal influence for tackling invasive weeds. Biodynamic farmers often schedule weeding activities during Mars's active phases, leveraging its energy to disrupt weed growth and promote healthier crop development.

3. **Saturn: Structure and Discipline:** Saturn, often referred to as the planet of structure and discipline, is highly relevant in biodynamic farming for activities that require meticulous planning and long-term focus. Its influence is particularly significant in soil management and foundational agricultural practices. The following are key areas associated with Saturn:

   - **Soil Preparation:** Saturn is considered ideal for activities like plowing, tilling, and composting. Its grounding energy is believed to promote soil stability and fertility, ensuring a strong foundation for future crop cycles.
   - **Tree Pruning and Maintenance:** Saturn's influence is also linked to the care and management of perennial plants, including fruit trees and shrubs. Farmers often time pruning activities to coincide with Saturnian phases, enhancing tree structure and promoting sustained health and productivity.
   - **Composting and Decomposition:** The planet's association with time and decay aligns with the processes of composting and organic matter breakdown. By timing composting activities with Saturn's cycles, biodynamic practitioners aim to accelerate decomposition and enrich the soil with vital nutrients.

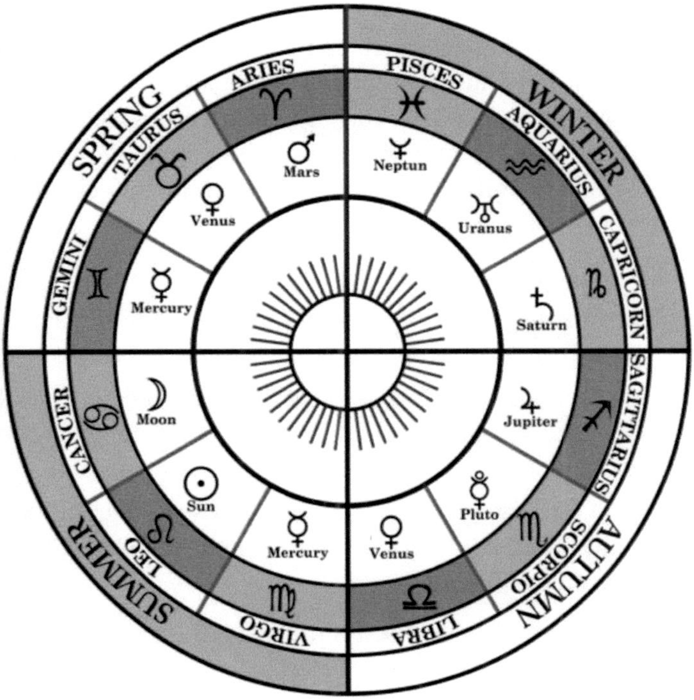

Fig. Planetary motion influence on seasons and Agricultural Timing

5.1 Integrating Planetary Influences in Practice

The integration of planetary influences into biodynamic farming requires careful observation and alignment of agricultural activities with celestial events. This practice is guided by biodynamic calendars, which provide detailed information on planetary movements, moon phases, and zodiac influences. By adhering to these calendars, farmers can:

- **Enhance Crop Quality:** Aligning planting and harvesting schedules with planetary cycles is believed to improve the taste, texture, and nutritional value of crops.
- **Optimize Resource Use:** Timing activities such as irrigation and fertilization with planetary energies ensures maximum efficiency and minimal waste.
- **Promote Ecological Balance:** The harmonization of farming practices with cosmic rhythms fosters a deeper connection with nature, promoting biodiversity and resilience within the farm ecosystem.

5.2 Scientific Perspectives and Future Research

While the principles of planetary influences in biodynamic farming are rooted in traditional wisdom, they have also sparked interest among researchers exploring the potential links between cosmic forces and agricultural outcomes. Studies investigating lunar and planetary effects on seed germination, plant growth, and soil microbiomes are gradually building a scientific framework to evaluate these practices.

Future research may focus on

- **Empirical Validation:** Rigorous experiments to assess the impact of specific planetary alignments on crop performance and soil health.
- **Integration with Technology:** Development of advanced tools and software to track planetary movements and provide real-time guidance for biodynamic practitioners.
- **Interdisciplinary Approaches:** Collaboration between agricultural scientists, astronomers, and ecologists to deepen understanding of the cosmic-agricultural connection.

6. Practical Applications of Lunar and Astrological Rhythms

In biodynamic farming, lunar and astrological rhythms are not merely theoretical concepts but are actively integrated into practical farming applications. This integration involves creating planting calendars, aligning agricultural practices with celestial cycles, and refining techniques based on observations. Below, we elaborate on the practical applications of these rhythms in agriculture, providing insights into how they are used to enhance farm management and productivity.

6.1 Creating a Planting Calendar

6.1.1 Understanding Lunar Phases

Lunar phases play a crucial role in determining the timing of various agricultural activities. The phases include the new moon, waxing moon, full moon, and waning moon. Each phase is believed to influence different aspects of plant growth and soil conditions:

- **New Moon:** Traditionally considered a time for planting root crops. The gravitational pull of the new moon is thought to encourage downward growth, making it an ideal period for establishing strong root systems.
- **Waxing Moon:** Spanning from the new moon to the full moon, this period is associated with rapid growth and development. It is often recommended for planting above-ground crops that benefit from increased light and energy.

- **Full Moon:** Associated with peak growth and vitality. This phase is seen as optimal for harvesting and weeding, as plants are believed to be at their most robust and nutrient-rich.
- **Waning Moon:** The period from the full moon to the new moon is considered suitable for soil preparation and planting root crops. The decreasing light is thought to promote inward growth and root development.

6.1.2 Incorporating Zodiac Signs

In addition to lunar phases, zodiac signs are used to further refine planting schedules. Each zodiac sign is linked to specific elements and qualities that influence plant growth:

- **Earth Signs (Taurus, Virgo, Capricorn):** Favorable for planting root crops and performing soil-related tasks. These signs are associated with stability and nourishment, promoting healthy root development and soil fertility.
- **Water Signs (Cancer, Scorpio, Pisces):** Ideal for planting leafy greens and crops that require ample moisture. These signs are linked to fluidity and hydration, supporting the growth of plants that thrive in wet conditions.
- **Air Signs (Gemini, Libra, Aquarius):** Suitable for planting flowering plants and crops requiring good air circulation. Air signs are associated with movement and growth, benefiting plants that need exposure to air and light.
- **Fire Signs (Aries, Leo, Sagittarius):** Favorable for planting crops that thrive in warmer conditions and require vigorous growth. These signs represent energy and vitality, supporting strong plant development.

6.1.3 Developing a Comprehensive Calendar

Farmers can create a detailed planting calendar that incorporates both lunar phases and zodiac signs. This calendar outlines optimal times for planting, harvesting, weeding, and soil preparation, ensuring that activities align with cosmic rhythms. The calendar can be customized based on specific crops, local climate conditions, and farm goals.

Fig. Planting calendar based on lunar phases

6.2 Aligning Agricultural Practices with Celestial Cycles

6.2.1 Soil Preparation and Fertility

Aligning soil preparation and fertility practices with lunar and astrological rhythms can enhance soil health and nutrient availability:

- **Soil Preparation During Waning Moon:** The waning moon is considered an appropriate time for soil preparation tasks, such as tilling and adding organic amendments. The decreasing light is believed to encourage the soil to absorb nutrients and prepare for future plant growth.
- **Application of Biodynamic Preparations:** Biodynamic preparations, such as horn manure and horn silica, can be applied based on lunar and zodiac cycles. For example, horn manure (Preparation 500) may be applied during the new moon to enhance soil structure and stimulate root growth.

6.2.2 Pest and Disease Management

Celestial rhythms are also used to guide pest and disease management practices:

- **Timing Pest Control with Planetary Cycles:** Certain planetary alignments, such as Mars' influence, are thought to impact pest activity. Farmers may schedule pest control measures and interventions based on these cycles to optimize effectiveness.
- **Weeding During Full Moon:** The full moon is considered a good time for weeding, as plants are believed to be at their most resilient and able to withstand disturbance. This timing helps prevent weed competition and supports healthy crop growth.

6.2.3 Harvesting and Post-Harvest Management

Harvesting and post-harvest management practices are optimized based on lunar and astrological rhythms:

- **Harvesting During Full Moon:** The full moon is regarded as the best time for harvesting, as crops are thought to be at their peak nutritional value. This timing can improve the quality and shelf life of harvested produce.
- **Post-Harvest Processing:** Post-harvest processing activities, such as drying and storage, can also be aligned with lunar cycles. For example, the waning moon may be used for storing and curing crops, as it is believed to support preservation and quality.

6.3 Refining Techniques Based on Observations

6.3.1 Monitoring Outcomes

Farmers practicing biodynamic agriculture continuously monitor the outcomes of their celestial-based practices. Observing plant growth, soil health, and overall farm productivity helps assess the effectiveness of lunar and astrological alignment:

- **Tracking Plant Performance:** Documenting plant growth patterns and yields in relation to lunar and astrological cycles provides valuable insights into the impact of cosmic rhythms on crop development.
- **Adjusting Practices:** Based on observations, farmers may adjust their planting schedules, soil management techniques, and pest control strategies to better align with cosmic influences and improve farm outcomes.

6.3.2 Adapting to Local Conditions

Farmers adapt celestial-based practices to suit local climate conditions and crop requirements. Regional variations, such as temperature, precipitation, and soil type, are considered when implementing lunar and astrological rhythms:

- **Customizing Practices:** Adapting celestial practices to local conditions ensures that farming techniques are effective and relevant to the specific environment. This customization enhances the overall success of celestial-based agricultural practices.
- **Continuous Learning:** Farmers engage in ongoing learning and experimentation to refine their techniques and better understand the interplay between cosmic rhythms and local conditions.

7. Benefits and Challenges

7.1 Benefits

7.1.1. Enhanced Plant Growth and Yield

- **Optimized Growth Phases:** Aligning planting activities with lunar phases can optimize growth conditions for different types of crops. For instance, planting root crops during the new moon and above-ground crops during the waxing moon can enhance growth patterns and potentially lead to higher yields.
- **Improved Nutrient Uptake:** By timing planting and cultivation activities with lunar cycles, farmers may improve nutrient availability and uptake. For example, the full moon's gravitational effects are thought to enhance nutrient absorption and plant vitality, potentially leading to healthier and more productive crops.

7.1.2 Better Soil Health and Fertility

- **Natural Soil Enhancement:** Celestial rhythms guide practices such as soil preparation and compost application. Applying biodynamic preparations and organic amendments during specific lunar phases can improve soil structure, fertility, and microbial activity.
- **Increased Organic Matter Recycling:** Timing compost application and soil amendments with lunar cycles can enhance the decomposition process and nutrient recycling. This practice helps build a rich, fertile soil environment conducive to healthy plant growth.

7.1.3 Effective Pest and Disease Management

- **Timing Pest Control:** Celestial rhythms are used to time pest control measures based on planetary cycles. For example, Mars' influence is believed to impact pest activity, allowing farmers to schedule interventions more effectively and reduce pest damage.

- **Reduced Pest Pressure:** By aligning weeding and pest management activities with lunar phases, farmers can minimize pest pressure and disease incidence. The full moon, for instance, is considered ideal for weeding when plants are more resilient and able to handle disturbances.

7.1.4 Sustainable Farming Practices

- **Holistic Approach:** Biodynamic farming's integration of cosmic rhythms promotes a holistic approach to agriculture, emphasizing the interconnectedness of soil, plants, animals, and celestial forces. This perspective encourages sustainable practices that support long-term farm health.
- **Reduced Chemical Inputs:** By optimizing planting and cultivation practices through celestial rhythms, farmers may reduce the need for synthetic fertilizers and pesticides, leading to more environmentally friendly and sustainable farming systems.

7.1.5 Increased Resilience and Adaptability

- **Enhanced Farm Resilience:** Aligning agricultural practices with cosmic rhythms may increase farm resilience to environmental stressors, pests, and diseases. By working with natural cycles, farmers can create a more balanced and adaptable farm ecosystem.
- **Adaptive Techniques:** Continuous observation and adjustment of celestial-based practices allow farmers to adapt to changing conditions and refine their approaches for better outcomes.

7.2 Challenges

7.2.1 Lack of Scientific Validation

- **Limited Research:** The influence of lunar and astrological rhythms on agriculture lacks robust scientific validation. Research supporting these practices is often anecdotal or based on traditional knowledge rather than rigorous scientific studies.
- **Skepticism and Acceptance:** The scientific community and mainstream agricultural practices may be skeptical of celestial-based methods, leading to challenges in gaining widespread acceptance and integration of these approaches.

7.2.2 Complexity and Labor Intensity

- **Complex Scheduling:** Integrating lunar and astrological rhythms into agricultural practices requires detailed planning and scheduling. Creating and adhering to planting calendars based on celestial cycles can be complex and time-consuming.
- **Increased Labor:** Aligning farming activities with cosmic rhythms may require additional labor and resources. For instance, precise timing of planting, weeding, and harvesting can be labor-intensive and demand careful coordination.

7.2.3 Practical Implementation Difficulties

- **Adaptation to Local Conditions:** Celestial-based practices may need to be adapted to local climate conditions and crop requirements. Implementing these practices effectively in diverse environments can be challenging and may require adjustments based on regional factors.
- **Resource Constraints:** Farmers with limited resources may face difficulties in adopting and maintaining celestial-based practices. Access to necessary materials, such as biodynamic preparations, and the ability to adhere to specific timings can be constraints.

7.2.4 Unpredictable Outcomes

- **Variability in Results:** The effectiveness of celestial-based practices can vary depending on factors such as weather conditions, soil type, and crop species. This variability may lead to unpredictable outcomes and challenges in achieving consistent results.
- **Observation and Adjustment:** Continuous monitoring and adjustment are required to assess the impact of celestial rhythms on farming practices. Farmers must be prepared to adapt their techniques based on observations and feedback, which can be a challenging and ongoing process.

7.2.5 Integration with Conventional Practices

- **Balancing Approaches:** Integrating celestial rhythms with conventional agricultural practices can be challenging. Farmers must find a balance between traditional methods and modern techniques to achieve optimal results while addressing potential conflicts or inconsistencies.
- **Educational and Training Needs:** Implementing celestial-based practices may require additional education and training for farmers. Access to resources, workshops, and support for understanding and applying these principles is essential for successful integration.

Conclusion

The influence of lunar and astrological rhythms on planting and cultivation practices represents a unique approach to agriculture that integrates cosmic and earthly processes. Biodynamic farming leverages these rhythms to guide planting, soil management, and crop cultivation, reflecting a holistic view of agriculture. While challenges exist in terms of scientific validation and practical implementation, the potential benefits of enhanced plant growth, improved soil health, and increased resilience underscore the value of exploring cosmic rhythms in farming. By understanding and applying these principles, farmers can optimize their cultivation practices, contributing to more sustainable and productive agricultural systems.

References

Jovchelevich, P. (2021). Astronomical Rhythms in Biodynamic Agriculture: A Brazilian Case Study on the Yield and Quality of Daucus carota L. under Biodynamic Management. In Subtle Agroecologies (pp. 123-130). CRC Press.

Kollerstrom, N., & Staudenmaier, G. (2001). Evidence for lunar-sidereal rhythms in crop yield: a review. Biological agriculture & horticulture, 19(3), 247-259.

Mayoral, O., Solbes, J., Cantó, J., & Pina, T. (2020). What has been thought and taught on the lunar influence on plants in agriculture? Perspective from physics and biology. Agronomy, 10(7), 955.

Spiess, H. (2000). Lunar rhythms and plants. BIODYNAMICS, 19-22.

Zürcher, E. (2006, October). Chronobiology of trees: Synthesis of traditional phytopractices and scientific research as tools of future forestry. In International Conference on Endogenous Development and Bio-cultural Diversity (pp. 3-5).

6

Planting Calendars, Guides, Cropping Systems in Biodynamic Farming

Abstract

Biodynamic farming is a holistic agricultural approach that integrates celestial rhythms, including lunar phases and astrological cycles, into farming practices to enhance soil health, plant growth, and farm productivity. Central to this practice are planting calendars and guides, which help farmers align their activities with cosmic rhythms. Additionally, cropping systems in biodynamic farming are designed to optimize resource use, improve soil fertility, and promote ecological balance. This article explores the principles and practical applications of planting calendars and guides in biodynamic farming, examines various cropping systems employed in biodynamic agriculture, and discusses their impact on farm management and sustainability. By understanding these elements, farmers can effectively incorporate biodynamic practices into their agricultural systems, contributing to more resilient and productive farming operations.

Introduction

In biodynamic farming, planting calendars, guides, and cropping systems form the backbone of a methodical approach to cultivating the land in harmony with natural and cosmic rhythms. These tools and practices are not just aids to scheduling agricultural tasks; they embody the essence of biodynamic principles by integrating spiritual, ecological, and cosmic insights into the daily operations of a farm. Unlike conventional farming, which often prioritizes efficiency and output, biodynamic farming places equal importance on the timing and sequence of agricultural activities, believing that the right alignment with natural forces can significantly enhance the vitality and productivity of the farm.

Planting calendars in biodynamic farming are meticulously crafted each year, drawing on the positions of the moon, planets, and constellations. These calendars provide detailed guidance on the optimal times for sowing seeds, transplanting, pruning, and harvesting. They are based on the understanding that different phases of the moon and specific astrological alignments influence

DOI: 10.1201/9781042015597-6

various aspects of plant growth, such as root development, leaf formation, flowering, and fruiting. Farmers who follow these calendars aim to harness the cosmic rhythms to improve the health and yield of their crops, ensuring that every action taken on the farm is in sync with the natural energies at play.

Guides accompanying these calendars offer practical advice on how to implement biodynamic principles throughout the farming process. They cover a wide range of topics, from soil preparation and composting to pest management and crop rotation. These guides often emphasize the importance of observation, encouraging farmers to develop a keen awareness of their environment and to adjust their practices according to the specific conditions of their land. This adaptive approach is key to the success of biodynamic farming, as it allows for a dynamic interaction between the farmer and the farm ecosystem.

Cropping systems in biodynamic farming are designed to create a self-sustaining and diverse agricultural environment. Crop rotation, intercropping, and polyculture are common practices that help maintain soil fertility, control pests and diseases, and promote biodiversity. These systems are carefully planned to ensure that each crop complements the others, creating a balanced and resilient ecosystem. Biodynamic farmers often choose crops based on their compatibility with the farm's specific microclimate and soil conditions, as well as their alignment with the biodynamic calendar.

The integration of planting calendars, guides, and cropping systems in biodynamic farming represents a holistic approach to agriculture that goes beyond conventional practices. It emphasizes the importance of timing, diversity, and ecological balance, all while maintaining a deep connection to the spiritual and cosmic dimensions of farming. This section will explore how these tools and systems are used to create a thriving, sustainable farm that nurtures both the land and the farmer, offering insights into the unique and complex world of biodynamic agriculture.

1. Planting Calendars in Biodynamic Farming

1.1 Concept and Importance

Planting calendars in biodynamic farming are essential tools that help farmers schedule planting, harvesting, and other agricultural activities according to cosmic rhythms. These calendars are based on lunar phases, zodiac signs, and planetary movements, reflecting the belief that celestial cycles influence plant growth, soil health, and overall farm vitality.

- **Lunar Phases:** Planting calendars incorporate the four primary lunar phases—new moon, waxing moon, full moon, and waning moon. Each phase is associated with different aspects of plant growth and soil conditions, guiding optimal times for planting various crops.

- **Zodiac Signs:** The twelve zodiac signs are also used to refine planting schedules. Each sign is linked to specific elements and qualities that influence plant growth, such as earth, water, air, and fire.
- **Planetary Cycles:** Planetary movements, such as those of Venus, Mars, and Saturn, are considered in creating planting calendars. These cycles are believed to impact aspects of plant health, soil fertility, and pest management.

1.2 Creating a Planting Calendar

- **Gathering Data:** Developing a planting calendar involves gathering data on lunar phases, zodiac signs, and planetary movements for the relevant growing season. This data is used to determine the best times for planting, harvesting, and other farm activities.
- **Customizing for Local Conditions:** Planting calendars should be customized to local climate conditions, soil types, and crop requirements. Regional variations, such as temperature and precipitation, are considered to ensure that the calendar aligns with specific environmental factors.
- **Integrating with Farm Practices:** The planting calendar is integrated with other farm practices, such as soil preparation, irrigation, and pest management. This integration helps ensure that all activities are coordinated and optimized according to cosmic rhythms.
- **Continuous Monitoring and Adjustment:** Farmers must continuously monitor the outcomes of their planting calendar and adjust practices based on observations and feedback. This ongoing process helps refine the calendar and improve its effectiveness.

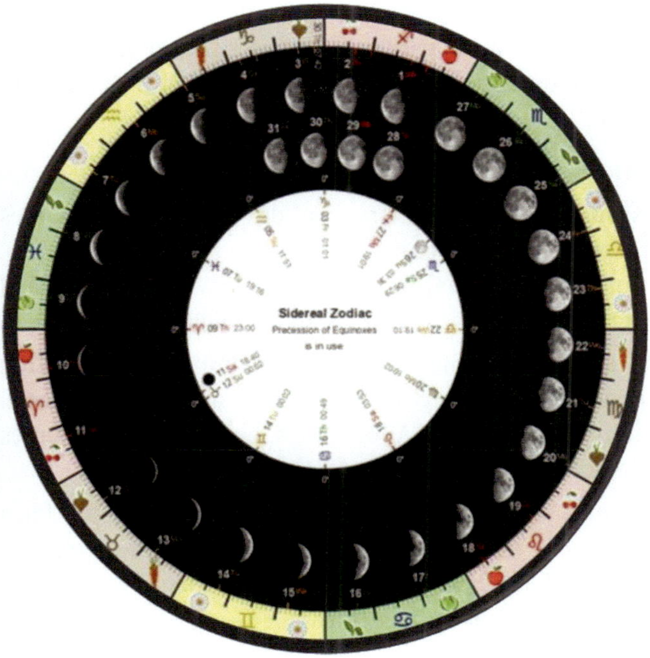

Fig. Sample of Planting Calendar based on lunar phases

2. Guides for Biodynamic Farming

2.1 Purpose and Use

Guides in biodynamic farming provide practical instructions and recommendations for implementing celestial-based practices. These guides cover various aspects of farm management, including soil preparation, planting, pest control, and harvesting.

- **Soil Preparation Guides:** These guides offer recommendations for preparing soil based on lunar phases and zodiac signs. They include information on tilling, compost application, and soil amendments to enhance soil fertility and structure.
- **Planting Guides:** Planting guides provide specific instructions for planting different types of crops according to cosmic rhythms. They include recommendations for timing, depth, and spacing based on lunar phases and zodiac signs.
- **Pest and Disease Management Guides:** These guides offer strategies for managing pests and diseases based on planetary cycles and lunar phases. They include recommendations for timing interventions, using biodynamic preparations, and promoting natural pest control methods.

- **Harvesting Guides:** Harvesting guides provide recommendations for timing harvests based on lunar phases and cosmic rhythms. They include information on optimal harvesting times to ensure peak quality and nutritional value.

2.2 Practical Application

- **Developing Custom Guides:** Farmers can develop custom guides tailored to their specific farm conditions, crops, and local climate. Custom guides integrate celestial-based practices with conventional farm management techniques to create a comprehensive approach.
- **Training and Education:** Farmers may seek training and education on biodynamic practices to effectively use guides and apply celestial rhythms in their farming operations. Workshops, courses, and resources are available to support farmers in implementing these practices.
- **Record-Keeping and Evaluation:** Maintaining records of farm activities, outcomes, and observations is essential for evaluating the effectiveness of guides. Farmers can use this information to refine their practices and improve their guides over time.

3. Cropping Systems in Biodynamic Farming

3.1 Principles of Biodynamic Cropping Systems

Biodynamic cropping systems are designed to optimize resource use, improve soil fertility, and promote ecological balance. These systems incorporate celestial rhythms, ecological principles, and holistic practices to create sustainable farming environments.

- **Diverse Crop Rotations:** Biodynamic cropping systems emphasize diverse crop rotations to enhance soil health and prevent pest and disease build-up. Rotating crops helps maintain soil fertility, reduce nutrient depletion, and promote biodiversity.
- **Companion Planting:** Companion planting involves growing complementary plants together to enhance growth and deter pests. Biodynamic practices incorporate celestial rhythms to guide companion planting decisions, optimizing plant interactions and benefits.
- **Cover Cropping:** Cover crops are used to improve soil health, prevent erosion, and add organic matter. Biodynamic cropping systems include guidelines for selecting and managing cover crops based on lunar phases and crop requirements.
- **Integrated Pest Management (IPM):** Biodynamic systems employ integrated pest management strategies that align with cosmic rhythms.

IPM practices include using natural predators, biodynamic preparations, and timing interventions based on planetary cycles.

3.2 Implementing Biodynamic Cropping Systems

- **Planning and Design:** Designing a biodynamic cropping system involves planning crop rotations, companion planting schemes, and cover cropping strategies. Farmers use planting calendars and guides to coordinate these practices with cosmic rhythms.
- **Monitoring and Evaluation:** Continuous monitoring of soil health, plant growth, and pest levels is essential for evaluating the effectiveness of biodynamic cropping systems. Farmers use observations and data to adjust practices and improve system performance.
- **Adapting to Local Conditions:** Biodynamic cropping systems should be adapted to local climate conditions, soil types, and crop requirements. Farmers may need to modify practices based on regional factors to achieve optimal results.
- **Community and Support:** Engaging with the biodynamic farming community and seeking support from experts can enhance the implementation of biodynamic cropping systems. Collaboration and knowledge-sharing contribute to successful adoption and ongoing improvement.

4. Benefits and Challenges of Planting Calendars, Guides, and Cropping Systems

The integration of planting calendars, guides, and cropping systems in biodynamic farming represents a unique approach to agriculture that harmonizes farm practices with cosmic rhythms. This approach aims to optimize soil health, plant growth, and farm productivity through careful alignment with lunar phases, zodiac signs, and planetary movements. While this method offers several benefits, it also presents specific challenges that need to be addressed for effective implementation.

4.1 Benefits

4.1.1 Enhanced Crop Productivity and Quality

- **Optimized Planting Timing:** Planting calendars based on lunar phases and zodiac signs help farmers schedule planting and harvesting activities at times that are believed to optimize plant growth. For instance, planting during the waxing moon can promote vigorous above-ground growth, potentially leading to higher yields and better-quality crops.

- **Improved Soil Fertility:** Guides for soil preparation and amendments, when aligned with cosmic rhythms, support better soil health. Practices such as applying compost or biodynamic preparations at the right times can enhance soil structure and fertility, leading to healthier plants and improved productivity.
- **Increased Nutrient Availability:** Planting and soil management activities synchronized with lunar cycles are thought to enhance nutrient availability and uptake. This can lead to crops that are more nutritious and resilient, benefiting both the environment and consumer health.

4.1.2 Sustainable Farming Practices

- **Reduced Chemical Inputs:** By optimizing natural processes through planting calendars and biodynamic practices, farmers may reduce reliance on synthetic fertilizers and pesticides. This aligns with the principles of sustainability and environmental stewardship, promoting healthier ecosystems.
- **Holistic Approach:** Biodynamic cropping systems emphasize a holistic view of the farm, integrating crop rotations, companion planting, and cover cropping. This approach fosters ecological balance and supports long-term soil health, contributing to sustainable farming practices.
- **Enhanced Biodiversity:** Diverse cropping systems and companion planting promote biodiversity on the farm. This diversity helps create balanced ecosystems, reduces pest pressures, and supports beneficial organisms, enhancing overall farm resilience.

4.1.3 Improved Farm Resilience and Adaptability

- **Resilient Cropping Systems:** Biodynamic practices, including diverse crop rotations and integrated pest management, contribute to resilient cropping systems. These systems can better withstand pests, diseases, and environmental stressors, supporting farm sustainability.
- **Adaptation to Environmental Changes:** Planting calendars and guides help farmers adapt to environmental changes by aligning practices with cosmic rhythms. This adaptability can improve farm performance and productivity in varying conditions.
- **Community Support and Knowledge Sharing:** Engaging with the biodynamic farming community and accessing guides and resources fosters knowledge sharing and support. This collaboration enhances farmers' ability to implement effective practices and address challenges.

4.1.4 Enhanced Farm Management and Planning

- **Structured Approach:** Planting calendars and guides provide a structured approach to farm management, helping farmers plan and execute activities more effectively. This organization can lead to better resource use and more efficient farm operations.
- **Data-Driven Decisions:** The use of planting calendars and guides allows farmers to make data-driven decisions based on cosmic rhythms. This informed approach can improve farm management and decision-making, leading to better outcomes.

4.2 Challenges

4.2.1 Complexity and Learning Curve

- **Understanding Cosmic Rhythms:** Implementing planting calendars and guides requires a deep understanding of lunar phases, zodiac signs, and planetary movements. This complexity can be challenging for farmers new to biodynamic practices and may require significant time and effort to master.
- **Developing Custom Calendars:** Creating and customizing planting calendars and guides to fit specific farm conditions and crop requirements can be intricate. Farmers must consider local climate, soil types, and crop needs, which adds to the complexity of implementation.

4.2.2 Limited Scientific Validation

- **Lack of Empirical Evidence:** The scientific validation of celestial-based practices in biodynamic farming is limited. Research supporting the effectiveness of lunar phases and zodiac signs in agriculture is often anecdotal or based on traditional knowledge rather than rigorous scientific studies.
- **Skepticism and Acceptance:** The scientific community and conventional agriculture may be skeptical of biodynamic practices, making it challenging for farmers to gain acceptance and support for these methods.

4.2.3 Resource Requirements and Practical Implementation

- **Additional Resources:** Implementing biodynamic planting calendars and guides may require additional resources, such as specialized tools, materials, and training. Farmers must balance these resource needs with their operational capacity and budget constraints.

- **Labor Intensity:** The need for precise timing and scheduling can increase labor intensity. Farmers may face challenges in coordinating activities according to cosmic rhythms while managing day-to-day farm operations.

4.2.4 Variability and Unpredictable Outcomes

- **Inconsistent Results:** The effectiveness of celestial-based practices can vary depending on factors such as weather conditions, soil type, and crop species. This variability may lead to inconsistent results and challenges in achieving desired outcomes.
- **Adaptation and Adjustment:** Continuous monitoring and adjustment of practices based on observations and feedback are required. This ongoing process can be demanding and may require farmers to be flexible and adaptive in their approach.

4.2.5 Integration with Conventional Practices

- **Balancing Approaches:** Integrating biodynamic practices with conventional farming methods can be challenging. Farmers must find a balance between traditional techniques and modern approaches to optimize farm management and productivity.
- **Educational and Training Needs:** Implementing biodynamic practices effectively requires education and training. Access to resources, workshops, and expert guidance is essential for successful integration and ongoing improvement.

Conclusion

Planting calendars, guides, and cropping systems play vital roles in biodynamic farming, offering frameworks for aligning agricultural practices with cosmic rhythms and promoting sustainable farm management. By understanding and implementing these elements, farmers can enhance soil health, optimize plant growth, and improve overall farm productivity. While challenges such as complexity, limited scientific validation, and resource requirements exist, the potential benefits of biodynamic practices, including improved farm resilience and sustainable practices, underscore their value in creating more effective and ecological farming systems. Through continuous learning, adaptation, and refinement, farmers can successfully integrate biodynamic principles into their agricultural operations, contributing to a more resilient and productive farming future.

References

Beluhova-Uzunova, R., & Atanasov, D. (2019). Biodynamic agriculture-old traditions and modern practices. Trakia Journal of Sciences, 17(1), 530-536.

Jayasree, P. (2005). Biodynamic practices in chilli (Doctoral dissertation, Department of Agronomy, College of Agriculture, Vellayani).

Koepf, H. H. (2005). The Biodynamic Farm: Agriculture in Service of the Earth and Humanity. SteinerBooks.

Ningombi, D., Dafouti, A., Nagpal, P., Nanda, H. B., Banerjee, S., & Kaushal, S. (2023). Dynamic use of Cow Pat Pit, Biodynamic Preparations and Peppering in Organic Farming. International Journal of Plant & Soil Science, 35(10), 157-167.

Ponzio, C., Gangatharan, R., & Neri, D. (2013). Organic and biodynamic agriculture: a review in relation to sustainability. International Journal of Plant & Soil Science, 2(1), 95-110.

Ravisankar, N., Panwar, A. S., Prasad, K., Kumar, V., & Bhaskar, S. (2017). Organic farming crop production guide. Network Project on Organic Farming, ICAR-Indian Institute of Farming Systems Research, Modipuram, Meerut-250, 110, 586.

7

Livestock Integration in Biodynamic Farming

Abstract

Biodynamic farming is an advanced agricultural approach that incorporates ecological, ethical, and holistic practices to create a self-sustaining ecosystem. A key component of biodynamic farming is livestock integration, which plays a crucial role in enhancing soil fertility, biodiversity, and farm productivity. This article explores the principles and practices of livestock integration in biodynamic farming, examining the benefits, challenges, and practical applications. By understanding the symbiotic relationships between animals, plants, and soil, farmers can create more resilient and sustainable farming systems. This comprehensive analysis provides insights into the methods and outcomes of integrating livestock into biodynamic farms, offering a roadmap for farmers seeking to adopt or enhance this practice.

Introduction

Livestock integration is a cornerstone of biodynamic farming, reflecting the holistic philosophy that underpins this agricultural approach. In biodynamic systems, animals are not merely an addition to the farm; they are integral components of a self-sustaining and balanced ecosystem. This integration goes beyond the traditional role of livestock in agriculture, emphasizing a symbiotic relationship between animals, crops, and the land itself. By incorporating livestock into the farm, biodynamic practices seek to enhance soil fertility, promote biodiversity, and create a closed-loop system where all elements work in harmony to support each other.

In biodynamic farming, livestock are considered vital contributors to the health and vitality of the farm. Their manure, enriched through specific biodynamic preparations, is used to create compost that revitalizes the soil, fostering a fertile environment for crops. This compost is more than just a fertilizer; it is seen as a living substance that enhances the soil's life forces, contributing to the overall resilience and productivity of the farm. The presence of animals also encourages a diversity of microorganisms and promotes the natural breakdown of organic matter, further enriching the soil. Beyond their role in soil health,

DOI: 10.1201/9781042015597-7

livestock in biodynamic farming help maintain the farm's ecological balance. Grazing animals, such as cows, sheep, and goats, contribute to the natural control of weeds and help cycle nutrients through the ecosystem. Their grazing patterns are carefully managed to ensure that they support the growth of diverse plant species, preventing overgrazing and promoting pasture regeneration. This approach not only sustains the land but also enhances the health and well-being of the animals, as they are raised in a natural and nurturing environment.

The integration of livestock in biodynamic farming also aligns with the broader principles of animal welfare and ethical farming. Biodynamic farmers prioritize the well-being of their animals, ensuring they are raised in conditions that respect their natural behaviors and life cycles. Animals are typically allowed to roam freely, graze on organic pastures, and interact with their environment in a way that promotes their health and vitality. This ethical treatment of animals is seen as essential to maintaining the overall harmony of the farm and is believed to result in healthier, more nutritious food products. Moreover, livestock integration in biodynamic farming contributes to the farm's economic sustainability. By diversifying farm outputs, such as meat, dairy, wool, and eggs, farmers can create multiple revenue streams, reducing their dependence on a single crop or product. This diversification also provides resilience against market fluctuations and environmental challenges, ensuring the long-term viability of the farm.

This section will delve into the various ways livestock are integrated into biodynamic farming, exploring their contributions to soil health, ecological balance, and economic sustainability. It will also examine the ethical considerations involved in raising animals within this holistic farming system, highlighting how biodynamic principles foster a deep connection between the farmer, the animals, and the land.

1. Historical Context of Biodynamic Farming

Biodynamic farming emerged as a response to the industrialization of agriculture, which led to soil degradation, loss of biodiversity, and a decline in food quality. Steiner's lectures in 1924 laid the groundwork for biodynamic practices, emphasizing the interconnectedness of all farm components. Livestock integration was highlighted as essential for creating a closed nutrient loop, where animals contribute manure and other organic matter to the soil, fostering a dynamic and balanced ecosystem.

1.1 Principles of Livestock Integration

Livestock integration in biodynamic farming is based on several core principles:

1. **Holistic Management:** Viewing the farm as an interconnected whole where animals, plants, and soil support each other.
2. **Biodiversity Enhancement:** Promoting a diverse range of species to create a resilient ecosystem.
3. **Closed Nutrient Loop:** Utilizing animal manure and organic matter to enrich the soil without relying on external inputs.
4. **Ethical Treatment of Animals:** Ensuring humane and natural living conditions for livestock.
5. **Soil Health:** Enhancing soil fertility through the natural recycling of nutrients.

2. Benefits of Livestock Integration

Integrating livestock into biodynamic farming systems yields numerous ecological, economic, and social benefits. These benefits extend beyond immediate agricultural productivity to encompass broader environmental sustainability and community health. Below, we delve into the key benefits in greater detail:

2.1 Soil Fertility Enhancement

Livestock play a crucial role in maintaining and enhancing soil fertility, a fundamental aspect of sustainable farming.

- **Nutrient-Rich Manure:** Animal manure is a valuable source of essential nutrients such as nitrogen, phosphorus, and potassium. When properly composted and applied to fields, manure improves soil structure, enhances microbial activity, and boosts nutrient availability to plants.
- **Organic Matter Addition:** The addition of organic matter from manure increases soil organic carbon content, improving soil texture and water retention. This leads to better root development and higher crop yields.
- **Enhanced Microbial Activity:** Manure and compost introduce beneficial microorganisms into the soil, which help in decomposing organic matter, fixing nitrogen, and improving soil health.

2.2 Biodiversity Promotion

Livestock integration supports biodiversity, which is vital for ecosystem resilience and stability.

- **Diverse Plant Species:** Grazing animals promote a variety of plant species by preventing any one species from becoming dominant. This plant diversity supports a wider range of insects, birds, and other wildlife.
- **Habitat Creation:** Livestock can create diverse habitats within the farm, such as hedgerows, ponds, and grassy areas, which provide homes for different species.
- **Pollinator Support:** By maintaining diverse plant life, livestock integration indirectly supports pollinators like bees and butterflies, which are essential for crop pollination.

2.3 Natural Pest and Weed Control

Livestock can significantly reduce the need for chemical pest and weed control measures.

- **Grazing Management:** Rotational grazing systems can control weed populations by allowing animals to graze specific areas intensively before moving on. This reduces the need for herbicides.
- **Pest Predation:** Certain livestock, such as poultry, feed on insects and pests, reducing their populations and the need for chemical pesticides.
- **Disruption of Pest Life Cycles:** Grazing can disrupt the life cycles of pests, reducing their ability to reproduce and infest crops.

2.4 Efficient Nutrient Cycling

Livestock help create a closed-loop system where nutrients are efficiently recycled within the farm.

- **Manure Composting:** Manure is composted and returned to the soil, ensuring that nutrients are recycled and not lost through runoff or leaching.
- **Minimized External Inputs:** By relying on internal nutrient sources, farms can reduce their dependence on synthetic fertilizers and external inputs, leading to cost savings and reduced environmental impact.
- **Enhanced Soil Health:** Continuous nutrient cycling improves soil health over time, leading to more productive and sustainable farming systems.

2.5 Economic Viability

Integrating livestock can improve the economic viability of farms through diversification and additional revenue streams.

- **Product Diversity:** Livestock provide multiple products such as meat, milk, eggs, wool, and leather, offering farmers diverse income sources.

- **Value-Added Products:** Farmers can produce value-added products like cheese, yogurt, and processed meats, which can command higher prices in the market.
- **Market Opportunities:** The growing demand for organic and sustainably produced animal products can open new market opportunities for biodynamic farmers.

2.6 Climate Resilience

Livestock integration can enhance the climate resilience of farming systems.

- **Carbon Sequestration:** Grazing lands and well-managed pastures can sequester significant amounts of carbon, helping mitigate climate change.
- **Adaptive Management:** Diverse farming systems with integrated livestock are more adaptable to changing climate conditions, as they can better withstand extreme weather events and other climate-related stresses.
- **Reduced Greenhouse Gas Emissions:** Properly managed livestock systems can reduce greenhouse gas emissions through practices such as rotational grazing and manure management.

2.7 Social and Community Benefits

Integrating livestock into farming systems can also have positive social and community impacts.

- **Rural Employment:** Livestock farming creates employment opportunities in rural areas, helping to sustain local economies.
- **Community Engagement:** Livestock farms often serve as educational and community hubs, where people can learn about sustainable farming practices and connect with nature.
- **Food Security:** By producing a variety of foods, integrated farms can contribute to local food security and reduce dependence on external food sources.

3. Challenges of Livestock Integration

While integrating livestock into biodynamic farming offers numerous benefits, it also presents several challenges. These challenges must be carefully managed to ensure the success and sustainability of the farming system. Here, we explore these challenges in detail:

3.1 Management Complexity

The integration of livestock into a farming system adds layers of complexity that require careful coordination and management.

- **Coordinating Livestock and Crops:** Ensuring that the needs of both livestock and crops are met without causing harm to either can be challenging. This includes timing grazing periods to avoid crop damage and ensuring that crop residues are available as fodder when needed.
- **Balancing Nutrient Flows:** Properly managing manure application to avoid nutrient imbalances or environmental contamination requires careful planning and monitoring.
- **Grazing Management:** Implementing rotational grazing systems to prevent overgrazing and maintain pasture health demands meticulous planning and execution.

3.2 Resource Requirements

Integrating livestock necessitates additional resources that can strain the farm's capacity and finances.

- **Feed and Water:** Livestock require a consistent supply of feed and water, which can be challenging during periods of drought or when pasture availability is limited.
- **Shelter and Infrastructure:** Providing adequate shelter, fencing, and other infrastructure for livestock involves significant investment and ongoing maintenance.
- **Labor:** Managing livestock is labor-intensive, requiring skilled labor to handle animal care, health monitoring, and farm management tasks.

3.3 Disease Management

The presence of livestock introduces the risk of diseases, which can impact both animal and crop health.

- **Disease Transmission:** Livestock can be carriers of diseases that may affect other animals, crops, and even humans. Managing these risks requires vigilant health monitoring and biosecurity measures.
- **Veterinary Care:** Ensuring the health of livestock necessitates regular veterinary care, vaccinations, and disease prevention measures, which can be costly and time-consuming.
- **Parasite Control:** Livestock can harbor parasites that may affect their health and productivity. Effective parasite control measures are essential but can be challenging to implement sustainably.

3.4 Infrastructure Investment

Establishing the necessary infrastructure for livestock integration involves significant upfront costs.

- **Capital Expenditure:** Building barns, shelters, fencing, water systems, and manure management facilities requires substantial investment.
- **Ongoing Maintenance:** Infrastructure requires regular maintenance and upgrades to ensure it remains functional and safe for livestock.
- **Financing:** Securing financing for these investments can be challenging, especially for small-scale or new farmers.

3.5 Market Access

Developing markets for animal products can be challenging, particularly for biodynamic farms that may operate on a smaller scale.

- **Market Competition:** Competing with larger, industrial-scale producers can be difficult, especially in terms of pricing and market reach.
- **Certification and Branding:** Obtaining biodynamic certification and building a brand that communicates the value of biodynamic products requires effort and resources.
- **Distribution Channels:** Establishing reliable distribution channels for animal products, especially perishable goods, can be complex and costly.

3.6 Environmental Impact

While livestock integration has many environmental benefits, it also poses potential environmental risks that must be managed.

- **Soil Compaction:** Overgrazing or improper grazing management can lead to soil compaction, reducing soil health and productivity.
- **Water Pollution:** Runoff from manure and livestock areas can contaminate water sources if not properly managed.
- **Greenhouse Gas Emissions:** Livestock, particularly ruminants, produce methane, a potent greenhouse gas. Managing emissions through sustainable practices is crucial.

3.7 Ethical Considerations

Ensuring the ethical treatment of animals is a core principle of biodynamic farming, but it poses several challenges.

- **Animal Welfare:** Providing humane living conditions, adequate space, and proper care for livestock requires ongoing commitment and resources.

- **Slaughter Practices:** Ethical slaughter practices must be adhered to, which can be challenging to implement and monitor consistently.
- **Consumer Perceptions:** Meeting consumer expectations for ethical treatment and transparency in animal husbandry practices requires clear communication and adherence to high standards.

4. Practical Applications of Livestock Integration

Integrating livestock into biodynamic farming systems requires strategic planning and implementation to maximize benefits and address challenges. Here, we delve into the practical applications of livestock integration, detailing methods and approaches that can be adopted by farmers to create a harmonious and productive farming ecosystem.

4.1 Selecting Suitable Livestock

Choosing the right type of livestock is crucial for successful integration.

- **Adaptation to Local Conditions:** Select livestock breeds that are well-suited to the local climate, soil types, and available forage. For instance, hardy sheep breeds may be more suitable for hilly, rough terrain, while dairy cows may thrive in lush, grassy areas.
- **Complementary Roles:** Different livestock species can play complementary roles. For example, cows and sheep can graze together as they prefer different types of vegetation, leading to more effective pasture utilization.
- **Multi-Purpose Breeds:** Consider breeds that provide multiple products (e.g., meat, milk, wool) to enhance farm productivity and economic resilience.

4.2 Designing Rotational Grazing Systems

Rotational grazing is a key practice in livestock integration, promoting soil health and pasture productivity.

- **Paddock Division:** Divide grazing land into multiple paddocks and rotate livestock through them. This allows pastures to rest and recover, preventing overgrazing and soil degradation.
- **Grazing Intensity:** Adjust the number of animals and grazing duration based on pasture growth and recovery rates. High-intensity, short-duration grazing can stimulate pasture regrowth and improve biodiversity.
- **Pasture Management:** Monitor pasture conditions and implement practices like overseeding with legumes and grasses to maintain healthy, productive pastures.

4.3 Composting Manure

Effective manure management is essential for nutrient cycling and soil fertility.

- **Composting Methods:** Use aerobic composting methods to break down manure into a stable, nutrient-rich organic fertilizer. Turn compost piles regularly to maintain oxygen levels and speed up decomposition.
- **Application Timing:** Apply composted manure to fields at appropriate times (e.g., before planting) to maximize nutrient availability and minimize nutrient runoff.
- **Manure Storage:** Store manure in covered areas to prevent nutrient loss through leaching and runoff, and to reduce odors and fly infestations.

4.4 Integrating Crop and Livestock Production

Coordinating crop and livestock activities can enhance farm productivity and sustainability.

- **Cover Crops and Grazing:** Use cover crops as forage for livestock. Grazing cover crops can improve soil health by adding organic matter and enhancing soil structure.
- **Crop Residues:** Allow livestock to graze on crop residues after harvest. This not only provides feed for animals but also aids in residue decomposition and nutrient cycling.
- **Agroforestry Systems:** Integrate trees and shrubs with pasture and crops to create silvopastoral systems. These systems provide shade and shelter for livestock, improve biodiversity, and enhance carbon sequestration.

4.5 Monitoring and Adaptation

Continuous monitoring and adaptation are essential to ensure the health and productivity of the integrated system.

- **Soil Testing:** Regularly test soil for nutrient levels, pH, and organic matter content. Adjust manure and compost applications based on test results to maintain optimal soil fertility.
- **Animal Health Monitoring:** Keep a close watch on livestock health through regular check-ups, vaccinations, and parasite control. Healthy animals are more productive and contribute positively to the farm ecosystem.
- **Record Keeping:** Maintain detailed records of grazing patterns, manure applications, crop yields, and livestock health. Use this data to make informed management decisions and improve practices over time.

4.6 Infrastructure Development

Investing in the right infrastructure is crucial for successful livestock integration.

- **Fencing:** Install sturdy fencing to manage livestock movement and protect crops. Electric fencing can be an effective and flexible option for rotational grazing systems.
- **Shelter:** Provide adequate shelter for livestock to protect them from extreme weather conditions. Barns, sheds, and natural shelters like trees can offer protection and comfort.
- **Water Systems:** Ensure a reliable water supply for livestock, especially during dry periods. Use troughs, ponds, or streams with clean, accessible water sources.

4.7 Ethical and Humane Practices

Adhering to ethical and humane practices is a cornerstone of biodynamic farming.

- **Animal Welfare:** Ensure that livestock have adequate space, access to clean water, nutritious feed, and opportunities to exhibit natural behaviors. Avoid practices that cause undue stress or harm.
- **Humane Slaughter:** Follow humane slaughter practices, ensuring minimal stress and pain for the animals. This includes proper handling, transportation, and use of appropriate methods.
- **Transparency:** Be transparent with consumers about farming practices and animal welfare standards. This can build trust and support for biodynamic products.

4.8 Community and Educational Engagement

Engaging with the community and providing educational opportunities can enhance the impact of livestock integration.

- **Farm Tours and Workshops:** Offer farm tours and workshops to educate the public about biodynamic farming practices, livestock integration, and sustainable agriculture.
- **Collaborative Networks:** Join or form networks with other biodynamic farmers to share knowledge, resources, and best practices.
- **School Programs:** Partner with local schools to create educational programs that teach children about farming, ecology, and the importance of sustainable practices.

5. Case Studies

Several successful case studies illustrate the benefits of livestock integration in biodynamic farming:

5.1 Case Study 1: Polyface Farm, Virginia, USA

- Polyface Farm integrates multiple species of livestock in a rotational grazing system, enhancing soil fertility and farm productivity. The farm's holistic approach has become a model for sustainable agriculture.

5.2 Case Study 2: Sekem, Egypt

- Sekem combines biodynamic practices with livestock integration to transform desert land into fertile farmland. The farm's success demonstrates the potential of biodynamic farming in challenging environments.

5.3 Case Study 3: Huxhams Cross Farm, UK

- Huxhams Cross Farm integrates dairy cattle, sheep, and poultry into its biodynamic system. The farm emphasizes the ethical treatment of animals and the creation of a self-sustaining ecosystem.

6. Future Directions

The future of livestock integration in biodynamic farming holds promise as well as challenges. Continued research and innovation are needed to address issues such as climate change, resource scarcity, and market dynamics. Embracing new technologies and practices can enhance the resilience and sustainability of biodynamic farms.

Conclusion

Livestock integration is a cornerstone of biodynamic farming, offering numerous ecological, economic, and ethical benefits. By understanding and implementing the principles and practices of livestock integration, farmers can create more resilient and sustainable agricultural systems. This article has provided a comprehensive overview of the theoretical foundations, practical applications, benefits, and challenges of livestock integration in biodynamic farming, offering a valuable resource for farmers, researchers, and policymakers.

References

Muhie, S.H. (2023). Concepts, principles, and application of biodynamic farming: a review. Circular Economy and Sustainability, 3(1), 291-304.

Ponzio, C., Gangatharan, R., & Neri, D. (2013). Organic and biodynamic agriculture: a review in relation to sustainability. International Journal of Plant & Soil Science, 2(1), 95-110.

Santoni, M., Ferretti, L., Migliorini, P., Vazzana, C., & Pacini, G. C. (2022). A review of scientific research on biodynamic agriculture. Organic Agriculture, 12(3), 373-396.

Scialabba, N.E.H. (2022). Introduction to livestock systems. In Managing Healthy Livestock Production and Consumption (pp. 3-25). Academic Press.

8

Biodynamic Certification and Standards in Biodynamic Farming

Abstract

Biodynamic farming, a pioneering method of organic agriculture founded by Rudolf Steiner in the 1920s, emphasizes a holistic, ecological approach to farming. Central to this practice is the concept of viewing the farm as a self-sustaining organism, integrating plants, animals, soil, and cosmic rhythms into a unified system. Biodynamic certification and standards are essential for ensuring that farming practices adhere to these holistic principles, providing transparency and credibility to consumers. This article explores the origins, principles, processes, and implications of biodynamic certification. It discusses key certification bodies, the rigorous standards they enforce, and the benefits and challenges associated with certification. Through detailed analysis, this article aims to offer a comprehensive understanding of how biodynamic certification supports sustainable agriculture and impacts the broader agricultural landscape.

Introduction

Biodynamic farming, inspired by the work of Rudolf Steiner, represents a unique approach to agriculture that emphasizes ecological balance, soil fertility, and ethical treatment of animals. Unlike conventional or even organic farming, biodynamic farming integrates the farm as a living organism, closely aligned with cosmic and natural cycles. Biodynamic certification serves as a formal recognition that a farm adheres to these principles, ensuring that products meet specific standards and providing assurance to consumers about the integrity of the farming practices. This article delves into the intricacies of biodynamic certification, including its history, core principles, certification processes, standards, and the associated benefits and challenges.

1. Origins and Historical Context

Biodynamic farming emerged in the early 20th century during a period of significant transformation in agriculture. The industrial revolution had introduced new technologies and practices that dramatically altered farming methods, leading to increased use of synthetic fertilizers, pesticides, and

DOI: 10.1201/9781042015597-8

monocultures. While these advancements brought about short-term gains in productivity, they also resulted in declining soil health, reduced biodiversity, and growing concerns about the long-term sustainability of agricultural practices.

Amid this backdrop of industrialization and environmental degradation, Rudolf Steiner, an Austrian philosopher, social reformer, and educator, began to explore alternatives to conventional farming. Steiner's work, which spans a variety of fields including education (Waldorf schools), medicine, and spirituality, also extended into agriculture. His lectures and writings on farming laid the foundation for what would become known as biodynamic agriculture.

Rudolf Steiner's Agricultural Course

udolf Steiner's Agricultural Course, also known as the "Spiritual Foundations for the Renewal of Agriculture," is the seminal series of lectures that laid the groundwork for what we now know as biodynamic farming. Delivered in Koberwitz (now Kobierzyce, Poland) in June 1924, Steiner's lectures addressed concerns raised by farmers about the declining health of soil, crops, and livestock due to the increasing use of chemical fertilizers and industrial farming practices. Steiner's insights provided a holistic, spiritual, and ecological alternative to the prevailing agricultural practices of the time.

1.1 Background and Context

In the early 20th century, European agriculture was undergoing significant changes, with the introduction of synthetic fertilizers and pesticides. These changes were seen as beneficial for increasing crop yields, but many farmers began to notice a decline in soil fertility, plant vitality, and animal health. Concerned about these trends, a group of farmers and agriculturalists sought guidance from Rudolf Steiner, who was known for his work in anthroposophy—a spiritual science that integrates philosophy, science, and spirituality.

Responding to these concerns, Steiner delivered a series of eight lectures over ten days, where he outlined his vision for a new approach to farming that would restore the health of the land and its produce. His lectures emphasized the need for a farming system that was not only sustainable but also aligned with the spiritual and cosmic rhythms of the Earth.

2. Key Concepts and Principles

Steiner's Agricultural Course introduced several key concepts and principles that form the foundation of biodynamic farming:

2.1 The Farm as a Living Organism

One of Steiner's central ideas is that a farm should be viewed as a self-sustaining, living organism. In this view, the farm is an interconnected whole, with each part—soil, plants, animals, and humans—contributing to and supporting the health of the entire system. The farm should ideally produce everything it needs, including fertilizers, feed, and seeds, minimizing external inputs and creating a closed-loop system.

2.2 Cosmic and Spiritual Forces

Steiner emphasized the importance of cosmic and spiritual forces in agriculture. He believed that the health of plants, animals, and soil is influenced not only by physical factors like nutrients and water but also by cosmic rhythms and spiritual energies. According to Steiner, the positions of the moon, planets, and stars affect plant growth, and these influences should be considered when planning farming activities such as planting, cultivating, and harvesting.

2.3 Biodynamic Preparations

One of the most distinctive aspects of Steiner's teachings is the use of biodynamic preparations. These are specific substances made from herbs, minerals, and animal manures, prepared in a particular way and applied to the soil, compost, or plants. Steiner introduced nine key preparations (numbered 500–508) that are used to enhance soil fertility, stimulate plant growth, and harmonize the farm with cosmic forces

2.4 Composting

Steiner placed significant emphasis on the preparation and use of compost as a vital component of biodynamic farming. He advocated for the use of biodynamic compost preparations (BD 502-507), made from specific medicinal herbs like yarrow, chamomile, and nettle, which are added to compost piles to enhance the decomposition process and improve the quality of the compost. The resulting compost is seen as a vital source of fertility that enlivens the soil and contributes to the overall health of the farm organism.

2.5 Balancing the Four Elements

Steiner also taught that a successful farm must maintain a balance between the four classical elements—earth, water, air, and fire. Each element is associated with different aspects of the farm:

- **Earth:** Relates to soil and physical matter.
- **Water:** Associated with plant fluids and sap flow.
- **Air:** Corresponds to the atmosphere and gas exchanges in plants.

- **Fire:** Linked to warmth and light, which are essential for photosynthesis and growth.

Balancing these elements ensures that the farm remains in harmony with the natural world and cosmic influences.

3. The Role of the Farmer

In Steiner's view, the farmer plays a crucial role as the steward of the land. The farmer must develop an intuitive understanding of the farm organism and work in harmony with natural and cosmic rhythms. This requires not only practical farming skills but also a spiritual connection to the land and an awareness of the forces at work in the environment. The farmer is seen as a co-creator with nature, actively participating in the life processes of the farm.

3.1 Legacy and Impact

Rudolf Steiner's Agricultural Course had a profound impact on the development of biodynamic farming and continues to influence agricultural practices worldwide. The principles he outlined have been adopted and adapted by farmers across the globe, leading to the establishment of biodynamic farms, educational programs, and certification bodies like Demeter, which ensures that products are produced according to biodynamic standards.

Steiner's ideas have also contributed to the broader movement for sustainable and regenerative agriculture, inspiring a deeper understanding of the relationship between agriculture, ecology, and spirituality. Today, biodynamic farming is recognized as a pioneering approach to organic and sustainable farming, offering a model for farming that is both ecologically sound and spiritually enriching.

4. Development and Early Adoption

Following Steiner's lectures, the principles of biodynamic farming began to gain traction among a small group of farmers who were intrigued by the holistic and ecological approach. The first formal organization to promote and implement biodynamic practices was the Anthroposophical Society, which was established to support Steiner's educational and agricultural ideas.

In 1928, the Demeter Association was founded in Germany, becoming the world's first certifying body for biodynamic agriculture. The name "Demeter" was chosen to honor the Greek goddess of agriculture and fertility, symbolizing the farm's connection to natural and cosmic forces. Demeter's mission was to develop and enforce standards for biodynamic farming and to provide certification for farms and products that adhered to these standards.

Early adopters of biodynamic farming faced challenges in implementing Steiner's ideas due to a lack of formal guidance and resources. However, the growing interest in organic and holistic farming practices led to increased research and development in biodynamic methods. Pioneering farmers and researchers began to refine and expand upon Steiner's original concepts, integrating new scientific insights while staying true to the core principles of biodynamic agriculture.

5. Expansion and Global Recognition

As biodynamic farming gained recognition, its principles began to spread beyond Germany. In the mid-20th century, the movement expanded to other European countries, including Austria, Switzerland, and France. The establishment of additional certification bodies and farming organizations helped to standardize and formalize biodynamic practices.

The global recognition of biodynamic farming grew over the following decades, with the establishment of Demeter certification programs in various countries and regions. Today, biodynamic farming is practiced in many parts of the world, including North America, Australia, and Asia. The principles and practices of biodynamic agriculture continue to influence and inspire sustainable farming movements, contributing to the broader discourse on ecological and holistic approaches to agriculture.

5.1 Legacy and Impact

The birth of biodynamic farming marked a significant shift in agricultural thinking, challenging the dominance of industrial and chemical-intensive farming practices. Steiner's vision of a holistic, ecological approach to farming has had a lasting impact on the agricultural community and has inspired a diverse range of practices and organizations committed to sustainability and environmental stewardship.

Biodynamic farming's emphasis on soil health, biodiversity, and ethical treatment of animals has contributed to the development of organic farming standards and has influenced modern approaches to sustainable agriculture. The certification and standards established by organizations like Demeter have provided a framework for ensuring the integrity of biodynamic practices and for supporting farmers in their pursuit of holistic and sustainable agriculture.

6. Formation of Certification Bodies

In 1928, the Demeter Association was established in Germany as the first certifying body for biodynamic farming. The name "Demeter" is derived from the Greek goddess of agriculture, symbolizing fertility and growth.

The association's mission was to develop and enforce standards that reflect the biodynamic principles laid out by Steiner. Over time, Demeter expanded globally, establishing certification programs in various countries and regions.

In addition to Demeter, other certification bodies emerged, such as the Biodynamic Association in the UK and Biodyvin for biodynamic viticulture. These organizations work in collaboration with Demeter and have contributed to the global recognition of biodynamic farming practices.

6.1 Core Principles of Biodynamic Certification

Biodynamic certification is grounded in several core principles that guide the farming practices and standards:

1. **Holistic Management**
   - **Farm as an Organism:** Biodynamic farming views the entire farm as a self-sustaining organism. This perspective encourages farmers to consider the interactions between plants, animals, soil, and the broader ecosystem.
   - **Integrated Practices:** Farmers integrate various farming practices to promote ecological balance, such as rotational grazing, composting, and the use of biodynamic preparations.
2. **Biodiversity**
   - **Diverse Ecosystems:** Biodynamic farms emphasize the importance of biodiversity, promoting a variety of plant and animal species to create resilient ecosystems.
   - **Habitat Creation:** Farmers establish diverse habitats, including hedgerows, ponds, and flowering plants, to support wildlife and pollinators.
3. **Soil Fertility**
   - **Natural Preparations:** Biodynamic practices involve the use of natural preparations, such as compost made from specific plant and animal materials, to enhance soil fertility and health.
   - **Crop Rotation:** Implementing diverse crop rotations helps prevent soil depletion and pest build-up.
4. **Ethical Treatment of Animals**
   - **Humane Living Conditions:** Biodynamic standards require that animals have access to pasture, shelter, and opportunities to exhibit natural behaviors.
   - **Organic Feed:** Animals are fed organic feed that is free from synthetic additives and GMOs.

5. **Closed Nutrient Cycles**
   - **Recycling Nutrients:** Biodynamic farms aim to minimize external inputs by recycling nutrients within the farm. Manure and compost are used to return nutrients to the soil.
   - **Resource Efficiency:** Farms strive to be self-sufficient, reducing reliance on external resources and inputs.
6. **Cosmic Rhythms**
   - **Lunar and Cosmic Cycles:** Biodynamic practices align planting and harvesting activities with lunar and cosmic rhythms, believing that these cycles influence plant growth and health.
   - **Natural Preparations:** Special preparations, such as those made from cow horns and silica, are used to harmonize with cosmic forces and enhance soil and plant health.

7. Certification Bodies and Standards

7.1 Demeter Association

The Demeter Association is the most recognized certifying body for biodynamic farming. Its certification standards are widely accepted and serve as a benchmark for biodynamic practices worldwide. Key aspects of Demeter certification include:

- **Standards Development:** Demeter's standards are developed through a rigorous process involving scientific research, practical experience, and input from biodynamic practitioners.
- **Global Reach:** Demeter operates globally, with certification programs in numerous countries. Each region adapts the standards to local conditions while maintaining the core biodynamic principles.
- **Continuous Improvement:** Demeter regularly reviews and updates its standards to incorporate new research, address emerging issues, and enhance the effectiveness of biodynamic practices.

Fig. Biodynamic certification bodies

Other Certification Bodies

- **Biodynamic Association:** This non-profit organization promotes biodynamic farming and offers certification services in collaboration with Demeter. It supports farmers through education, research, and advocacy.
- **Biodyvin:** Specializing in biodynamic viticulture, Biodyvin certifies vineyards that adhere to biodynamic principles. It ensures that wine production aligns with the holistic practices of biodynamic farming.

8. Certification Process

Obtaining certification for biodynamic farming involves meeting stringent standards and adhering to practices that align with the holistic principles of the biodynamic approach. The certification process typically includes the following steps:

1. **Initial Assessment:** Farmers interested in biodynamic certification must first contact a recognized certifying body, such as Demeter International or its national affiliates. An initial assessment is conducted to evaluate the farm's current practices and its potential to meet certification standards.
2. **Transition Period:** Farms transitioning from conventional or organic methods to biodynamic farming usually undergo a conversion period of 2-3 years. During this time, farmers implement biodynamic practices, such as using biodynamic preparations, following lunar planting calendars, and fostering biodiversity. This period allows the farm ecosystem to adapt to the holistic approach.

3. **Implementation of Biodynamic Practices:** To qualify for certification, farms must demonstrate adherence to biodynamic principles, including:
    - Use of compost, biodynamic preparations, and natural inputs.
    - Maintenance of closed-loop nutrient cycles within the farm.
    - Diversified cropping systems and integrated livestock management.
    - Avoidance of synthetic fertilizers, pesticides, and genetically modified organisms (GMOs).
4. **Farm Inspection:** An on-site inspection is conducted by a certified inspector to verify compliance with biodynamic standards. The inspection includes a review of farm records, soil health, crop diversity, livestock integration, and the use of biodynamic preparations.
5. **Certification Approval:** Following a successful inspection, the certifying body reviews the findings and determines whether the farm meets all biodynamic criteria. Once approved, the farm is granted biodynamic certification, allowing its products to carry the Demeter label.
6. **Ongoing Compliance:** Certified farms must undergo annual inspections and maintain detailed records to ensure continued adherence to biodynamic standards. Any deviations from the standards can result in suspension or revocation of certification.

The certification process not only validates the farm's commitment to sustainable and holistic agriculture but also assures consumers of the authenticity and quality of biodynamic products.

9. Application and Inspection

The certification process involves several key steps to ensure that farms meet biodynamic standards:

1. **Application**
    - **Initial Submission:** Farmers interested in certification submit an application to the certifying body, detailing their farming practices, land, and livestock.
    - **Documentation:** The application includes documentation on soil management, crop rotation, animal welfare, and the use of biodynamic preparations.
2. **Initial Inspection**
    - **On-Site Assessment:** An inspector visits the farm to evaluate compliance with biodynamic standards. This assessment covers

various aspects, including soil health, crop management, animal welfare, and the use of biodynamic preparations.

- **Inspection Report:** The inspector prepares a detailed report, highlighting areas of compliance and any issues that need addressing.

3. **Compliance Review**

- **Evaluation:** The certifying body reviews the inspection report and determines whether the farm meets the necessary standards for certification.
- **Certification Decision:** Based on the review, the certifying body grants or denies certification. If granted, the farm receives a certificate that is valid for a specified period.

9.1 Ongoing Compliance

Certified farms must maintain ongoing compliance with biodynamic standards to retain their certification:

1. **Annual Inspections**
   - **Regular Checks:** Certified farms undergo annual inspections to ensure continued adherence to biodynamic standards. These inspections help identify and address any issues that may arise.
   - **Spot Checks:** Random spot checks may be conducted to verify compliance and prevent fraud.
2. **Record Keeping**
   - **Detailed Records:** Farmers must maintain detailed records of farming practices, inputs, and outputs. These records are used to demonstrate compliance during inspections.
   - **Transparency:** Proper documentation ensures transparency and accountability in biodynamic farming practices.
3. **Updates and Renewals**
   - **Periodic Reviews:** Certification is renewed periodically, requiring farms to update their practices and documentation as needed.
   - **Standard Updates:** Farms must stay informed about changes to certification standards and incorporate any new requirements.

9.2 Key Standards and Regulations

Biodynamic certification standards cover a wide range of practices and regulations to ensure that farms adhere to biodynamic principles:

Soil and Crop Management

- **Soil Health:** Biodynamic standards require the use of natural preparations to enhance soil fertility and structure. This includes composting with specific plant and animal materials and using preparations made from cow horns and silica.
- **Crop Rotation:** Farmers must implement diverse crop rotations to maintain soil health and prevent the build-up of pests and diseases.
- **Composting:** Creating and using compost is essential for recycling organic matter and nutrients back into the soil.

Animal Welfare

- **Humane Treatment:** Biodynamic standards mandate that animals have access to pasture, shelter, and space to exhibit natural behaviours. This ensures their physical and psychological well-being.
- **Organic Feed:** Animals must be fed organic feed that is free from synthetic additives, hormones, and GMOs.
- **Health and Care:** Regular veterinary care, vaccinations, and humane handling practices are required to maintain animal health and welfare.

Biodiversity and Ecosystem Management

- **Habitat Creation:** Biodynamic farms are encouraged to create and maintain diverse habitats for wildlife, including planting hedgerows, establishing ponds, and providing nesting sites for birds and insects.
- **Pollinator Support:** Farms must ensure that there is a diverse range of flowering plants to support pollinators such as bees and butterflies.
- **Ecosystem Balance:** Maintaining a balanced ecosystem with minimal external inputs is crucial for the sustainability of biodynamic farming systems.

Cosmic and Natural Cycles

- **Planting Calendar:** Biodynamic practices align planting and harvesting activities with lunar and cosmic rhythms. Farmers follow specific lunar phases and celestial events to optimize plant growth and health.
- **Natural Preparations:** Biodynamic preparations, such as those made from fermented plant materials and animal manures, are used to enhance soil and plant health and harmonize with cosmic forces.

10. Benefits of Biodynamic Certification

10.1 For Farmers

- **Market Differentiation:** Certification helps farmers differentiate their products in the market, often allowing them to command premium prices for biodynamic products.
- **Sustainable Practices:** Adhering to biodynamic standards promotes sustainable farming practices, leading to improved soil health, biodiversity, and farm resilience.
- **Community and Network:** Certified farmers join a global community of biodynamic practitioners, providing opportunities for knowledge sharing, collaboration, and support.

10.2 For Consumers

- **Quality Assurance:** Certification provides consumers with assurance that products are produced according to rigorous standards for sustainability, quality, and ethical practices.
- **Health Benefits:** Biodynamic products are often perceived as healthier due to the emphasis on natural and organic practices, free from synthetic chemicals and GMOs.
- **Environmental Impact:** Supporting biodynamic products contributes to environmental sustainability, promoting practices that enhance soil health, biodiversity, and ecological balance.

10.3 For the Environment

- **Soil Health:** Biodynamic practices improve soil structure, fertility, and carbon sequestration, contributing to long-term agricultural sustainability.
- **Biodiversity:** Certified farms support biodiversity by promoting diverse plant and animal species, creating habitats for wildlife, and enhancing ecosystem resilience.
- **Climate Resilience:** Biodynamic farms are better equipped to adapt to climate change due to their holistic and sustainable practices, which promote soil health and ecological balance.

DEMETER CERTIFIED BIODYNAMIC

DEMETER ASSOCIATION, INC.
hereby certifies that

MARIAN FARMS of Fresno, CA
Operated by
Gena Nonini
9835 West McKinley
Fresno, CA 93723

is a DEMETER CERTIFIED BIODYNAMIC® FARM
according to the Demeter US standards.

Scope of Certification: Crops
Certified Production: Per Attached Product Profile

Only the products listed on the attached Product Profile are eligible to carry the Demeter and Biodynamic certification marks. Rights to use of certification marks are contracted annually and are not transferable.

Effective Dates: January 1, 2020 to March 1, 2021
Issue Date: January 1, 2020
Certification ID#: BDF0147R0120

Authorized by:

Sarah Rhynalds, Member Service Coordinator

For further information, contact Demeter Association, Inc.
Phone: (541) 929-7148 Email: info@demeter-usa.org www.demeter-usa.org

Fig. Sample Biodynamic certificate of Marian biodynamic farm

11. Challenges of Biodynamic Certification

11.1 Cost and Accessibility

- **Certification Costs:** The cost of certification, including application fees, inspections, and compliance costs, can be prohibitive for small-scale farmers.
- **Access to Resources:** Farmers in remote or underserved areas may face challenges accessing the resources, training, and support needed for certification.

11.2 Compliance and Documentation

- **Record Keeping:** Maintaining detailed records and documentation can be burdensome, particularly for farmers new to biodynamic practices or those with limited administrative support.
- **Inspection Demands:** The frequency and thoroughness of inspections can be demanding, requiring farmers to invest significant time and effort into preparing for and responding to inspections.

11.3 Market Recognition

- **Consumer Awareness:** Limited consumer awareness and understanding of biodynamic certification can impact market demand and the ability to command premium prices.
- **Market Access:** Gaining access to markets that value and recognize biodynamic certification can be challenging, especially for new or smaller producers.

12. Case Studies

12.1 Case Study 1: Polyface Farm, Virginia, USA

Polyface Farm, operated by Joel Salatin, is a well-known example of biodynamic farming in the United States. The farm integrates biodynamic principles with innovative practices, such as rotational grazing and composting. Polyface Farm's success illustrates the benefits of biodynamic certification in enhancing soil health, farm productivity, and market differentiation.

12.2 Case Study 2: Sekem, Egypt

Sekem is a pioneering biodynamic farm located in the Egyptian desert. Founded by Ibrahim Abouleish, Sekem transformed arid land into fertile farmland using biodynamic practices. The farm's success in creating a thriving ecosystem and supporting local communities highlights the potential of biodynamic farming in challenging environments and the impact of certification on sustainable development.

12.3 Case Study 3: Huxhams Cross Farm, UK

Huxhams Cross Farm in the UK integrates dairy cattle, sheep, and poultry into its biodynamic system, adhering to Demeter certification standards. The farm's emphasis on ethical treatment of animals, soil fertility, and ecological balance provides a model for biodynamic certification in diverse agricultural settings. Huxhams Cross Farm demonstrates how certification supports sustainable practices and enhances farm resilience.

13. Future Directions

13.1 Evolving Standards

As biodynamic farming practices continue to evolve, certification standards must adapt to new challenges and opportunities. This includes incorporating advances in sustainable agriculture, addressing climate change impacts, and integrating new research findings.

13.2 Expanding Market Reach

Increasing consumer awareness and demand for biodynamic products will be crucial for expanding market reach. Efforts to educate consumers, develop effective marketing strategies, and collaborate with retailers and policymakers can help drive growth in the biodynamic market.

13.3 Supporting Small-Scale Farmers

Ensuring that certification is accessible and affordable for small-scale farmers is essential for the growth and diversity of biodynamic agriculture. This may involve providing financial support, resources, and training to help farmers achieve and maintain certification.

Conclusion

Biodynamic certification and standards are fundamental to ensuring the integrity and sustainability of biodynamic farming practices. By adhering to rigorous standards, certified farms promote soil health, biodiversity, and ethical treatment of animals, offering numerous benefits to farmers, consumers, and the environment. Despite the challenges, the future of biodynamic certification holds promise, with opportunities for growth, innovation, and increased market recognition. This comprehensive overview provides valuable insights into the principles, processes, and impacts of biodynamic certification, contributing to the advancement of sustainable agriculture and the broader agricultural community.

References

Beluhova-Uzunova, R., & Atanasov, D. (2019). Biodynamic agriculture-old traditions and modern practices. Trakia Journal of Sciences, 17(1), 530-536.

Muhie, S. H. (2023). Concepts, principles, and application of biodynamic farming: a review. Circular Economy and Sustainability, 3(1), 291-304.

Ponzio, C., Gangatharan, R., & Neri, D. (2013). Organic and biodynamic agriculture: a review in relation to sustainability. International Journal of Plant & Soil Science, 2(1), 95-110.

Santoni, M., Ferretti, L., Migliorini, P., Vazzana, C., & Pacini, G. C. (2022). A review of scientific research on biodynamic agriculture. Organic Agriculture, 12(3), 373-396.

9

Economic and Market Considerations of Biodynamic Farming

Abstract

Biodynamic farming, a holistic and ecological approach to agriculture founded by Rudolf Steiner, has gained prominence as a sustainable and ethical alternative to conventional farming methods. This method integrates ecological balance, soil fertility, and cosmic rhythms into farming practices. Despite its growing popularity, the economic viability and market dynamics of biodynamic farming present both opportunities and challenges. This article explores the economic aspects of biodynamic farming, including production costs, market demand, pricing strategies, and financial sustainability. It also examines the market trends, consumer preferences, and the role of certification in influencing economic outcomes. By analysing case studies and market data, this article provides a comprehensive overview of the economic and market considerations of biodynamic farming, offering insights into its viability and impact within the broader agricultural industry.

Introduction

Biodynamic farming represents a distinctive approach to agriculture that emphasizes ecological balance, soil health, and the integration of cosmic rhythms. Unlike conventional farming, which often relies on synthetic chemicals and monocultures, biodynamic farming aims to create a self-sustaining agricultural ecosystem. While the environmental and ethical benefits of biodynamic farming are well-documented, understanding its economic aspects is crucial for assessing its viability and impact on the agricultural market. This article delves into the economic and market considerations of biodynamic farming, exploring production costs, market dynamics, and financial sustainability.

1. Economic Aspects of Biodynamic Farming

1.1 Production Costs

Biodynamic farming involves several practices that can influence production costs:

DOI: 10.1201/9781042015597-9

1.1.1. Certification Costs

- **Initial Certification Fees:** Obtaining biodynamic certification requires an initial fee, which can vary based on the size of the farm and the certification body. This fee covers the application process, inspections, and documentation.
- **Ongoing Compliance Costs:** Maintaining certification involves annual fees, periodic inspections, and compliance with updated standards. These costs can add up over time, impacting the overall budget for biodynamic farms.

1.1.2 Labor Costs

- **Manual Labor:** Biodynamic farming often relies on manual labor for tasks such as planting, weeding, and harvesting. This can increase labor costs compared to conventional farming, which may use mechanized equipment.
- **Specialized Skills:** The implementation of biodynamic practices may require specialized skills and knowledge, necessitating additional training and potentially higher wages for skilled labor.

1.1.3 Input Costs

- **Biodynamic Preparations:** The use of biodynamic preparations, such as compost and herbal teas, involves additional costs for materials and production. These preparations are integral to soil fertility and plant health but can be more expensive than synthetic alternatives.
- **Organic Inputs:** Biodynamic farms use organic inputs and materials, which can be costlier than conventional inputs. This includes organic seeds, fertilizers, and pest control methods.

1.1.4 Infrastructure and Equipment

- **Farm Infrastructure:** Biodynamic farms may require investments in infrastructure, such as composting facilities, greenhouses, and animal housing. These investments are essential for implementing biodynamic practices effectively.
- **Equipment:** The need for specialized equipment, such as compost spreaders or manual cultivation tools, can add to the initial setup and maintenance costs.

2. Financial Sustainability

Assessing the financial sustainability of biodynamic farming involves analysing several factors:

2.1 Return on Investment (ROI)

- **Premium Pricing:** Biodynamic products often command premium prices in the market due to their perceived quality and sustainability. This can lead to higher returns for farms that successfully market their products.
- **Yield and Productivity:** The yields and productivity of biodynamic farms can vary compared to conventional farms. Factors such as soil health and biodiversity can influence crop yields and overall farm performance.

2.2 Cost-Benefit Analysis

- **Long-Term Benefits:** While initial costs may be higher, biodynamic farming can offer long-term benefits such as improved soil health, reduced reliance on external inputs, and increased resilience to pests and diseases.
- **Economic Resilience:** Biodynamic farms may experience greater economic resilience due to their emphasis on sustainability and self-sufficiency, which can mitigate the impacts of market fluctuations and external pressures.

2.3 Access to Financial Support

- **Grants and Subsidies:** Various grants, subsidies, and financial incentives are available to support biodynamic and organic farming practices. Access to these resources can help offset some of the costs associated with certification and implementation.
- **Loans and Investment:** Farmers may seek loans or investment opportunities to support the transition to biodynamic farming or to expand their operations. Financial institutions and investors may be attracted to the sustainable and ethical aspects of biodynamic agriculture.

3. Market Demand for Biodynamic Products

3.1 Consumer Preferences

3.1.1 Healt*h and Wellness Trends*

- **Perceived Health Benefits:** Consumers are increasingly interested in products perceived as healthier and more natural. Biodynamic products are often associated with higher quality and fewer synthetic chemicals, appealing to health-conscious consumers.
- **Organic and Ethical Choices:** The demand for organic and ethically produced food has been rising, with biodynamic products fitting into

this broader trend. Consumers value transparency, environmental stewardship, and animal welfare, which are key aspects of biodynamic farming.

3.1.2 Sustainability and Environmental Impact

- **Environmental Concerns:** Growing awareness of environmental issues and climate change has driven consumer preference for sustainable and eco-friendly products. Biodynamic farming's focus on soil health, biodiversity, and reduced chemical use aligns with these values.
- **Carbon Footprint:** Biodynamic farms often emphasize practices that reduce carbon footprints and enhance ecological balance, resonating with consumers who prioritize sustainability.

3.1.3 Market Segmentation

- **Niche Markets:** Biodynamic products often target niche markets where consumers are willing to pay a premium for high-quality, sustainably produced goods. These markets include specialty stores, farmer's markets, and premium grocery chains.
- **Mainstream Markets:** As awareness and demand for biodynamic products grow, there is potential for these products to enter mainstream markets. Expanding market reach requires effective marketing and distribution strategies.

4. Pricing Strategies

4.1 Premium Pricing

- **Market Positioning:** Biodynamic products are typically positioned as premium offerings, reflecting their higher production costs and the added value of sustainability and quality. This premium pricing strategy can enhance profitability for farms.
- **Consumer Willingness:** The willingness of consumers to pay higher prices for biodynamic products can vary based on factors such as product category, market trends, and economic conditions.

4.2 Pricing Models

- **Direct Sales:** Farms that sell directly to consumers through farm shops, online platforms, or subscription models can set prices that reflect the value of their products and maintain higher margins.
- **Retail and Wholesale:** Pricing strategies for retail and wholesale channels must consider factors such as distribution costs, retailer markups, and market competition.

4.3 Cost Management

- **Efficiency Improvements:** Biodynamic farms can focus on improving efficiency and reducing costs through practices such as crop rotation, composting, and integrated pest management. Efficient resource use can help balance production costs and pricing strategies.
- **Scaling Up:** Expanding farm operations and increasing production can help achieve economies of scale, potentially reducing per-unit costs and supporting competitive pricing.

5. Market Trends and Opportunities

5.1 Growth of Organic and Sustainable Markets

*5.1.1 Mar*ket Expansion

- **Rising Demand:** The market for organic and sustainable products has been expanding globally, driven by consumer awareness and preference for healthier and environmentally friendly options. Biodynamic farming is well-positioned to capitalize on this growth.
- **Product Innovation:** Innovations in product development, such as new biodynamic crops, value-added products, and unique offerings, can attract new customers and drive market growth.

5.1.2 Retail and Distribution Channels

- **Specialty Retailers:** Specialty retailers and natural food stores are key distribution channels for biodynamic products. Building relationships with these retailers and effectively communicating the value of biodynamic practices can enhance market reach.
- **Online Platforms:** E-commerce and online platforms offer opportunities for biodynamic farms to reach a broader audience and sell directly to consumers. Online sales channels can complement traditional retail strategies.

5.2 Consumer Education and Advocacy

5.2.1 Increasing Awareness

- **Educational Campaigns:** Educating consumers about the benefits and principles of biodynamic farming can drive demand and support for biodynamic products. Campaigns can focus on sustainability, health benefits, and the ethical aspects of biodynamic practices.
- **Certification Recognition:** Promoting the credibility and transparency of certification bodies, such as Demeter, can enhance consumer trust and confidence in biodynamic products.

5.2.2. Advocacy and Collaboration

- **Partnerships:** Collaborating with environmental organizations, industry associations, and advocacy groups can raise awareness and support for biodynamic farming. Partnerships can also facilitate knowledge sharing and resources for farmers.
- **Policy Support:** Advocating for policies and incentives that support biodynamic farming and sustainable agriculture can help
- create a favourable environment for growth and development.

6. Case Studies

6.1 Case Study 1: Rodale Institute, USA

The Rodale Institute, located in Kutztown, Pennsylvania, is a leading organization in the field of organic and biodynamic agriculture. Founded in 1947 by J.I. Rodale, the institute has been at the forefront of research, education, and advocacy for sustainable farming practices. It is renowned for its commitment to advancing organic and biodynamic farming through rigorous scientific research, hands-on demonstration, and farmer training.

Historical Context

The Rodale Institute's mission has been to promote and develop practices that enhance soil health, improve farm productivity, and support environmental sustainability. Initially focused on organic farming, the institute has expanded its research and outreach to include biodynamic agriculture as part of its broader commitment to sustainable farming practices.

In the early 2000s, the institute began incorporating biodynamic practices into its research programs, recognizing the potential of biodynamic methods to further enhance soil fertility, ecosystem health, and farm resilience.

Research and Demonstration

1. Biodynamic Farming Systems Trial

The Rodale Institute conducts long-term research trials comparing biodynamic farming systems with conventional and organic farming methods. The key objectives of these trials are to assess the performance of biodynamic practices in terms of soil health, crop yields, and environmental impact.

- **Soil Health:** Research has shown that biodynamic practices can significantly improve soil structure, organic matter content, and microbial activity. These improvements contribute to better soil fertility and reduced erosion.

- **Crop Yields:** While initial yields in biodynamic systems may be comparable to or slightly lower than conventional systems, long-term studies often reveal increased resilience to pests and diseases, leading to more stable yields over time.
- **Environmental Impact:** Biodynamic farming's focus on ecological balance and reduced chemical inputs has been associated with positive environmental outcomes, including enhanced biodiversity and improved water quality.

2. Educational Programs and Workshops

The Rodale Institute offers a range of educational programs and workshops focused on biodynamic farming. These programs are designed to provide farmers with practical knowledge and skills to implement biodynamic practices on their farms.

- **Farmer Training:** Workshops and training sessions cover topics such as biodynamic preparations, soil management, and crop rotation. The institute's hands-on approach helps farmers gain practical experience and build confidence in applying biodynamic methods.
- **Online Resources:** The institute provides online resources, including webinars, articles, and guides, to reach a wider audience and support farmers who may not be able to attend in-person events.

3. Partnerships and Collaborations

The Rodale Institute collaborates with other research institutions, agricultural organizations, and certification bodies to advance the understanding and adoption of biodynamic farming.

- **Research Partnerships:** Collaborations with universities and research institutions enable the institute to conduct interdisciplinary research and explore new innovations in biodynamic agriculture.
- **Certification and Advocacy:** The institute works with certification bodies such as Demeter to promote and support biodynamic certification. It also advocates for policies and practices that support sustainable agriculture.

Economic Considerations

1. Cost-Benefit Analysis

The Rodale Institute's research includes a detailed cost-benefit analysis of biodynamic farming practices. This analysis considers factors such as input costs, labor requirements, and market premiums.

- **Input Costs:** Biodynamic farming involves the use of specialized preparations and organic inputs, which can be more expensive than conventional alternatives. However, the long-term benefits of improved soil health and reduced need for external inputs can offset these costs.
- **Labor Requirements:** Biodynamic farming often requires more manual labor and specialized skills. While this can increase labor costs, the emphasis on soil health and ecological balance can lead to greater efficiency and reduced need for chemical inputs.
- **Market Premiums:** Biodynamic products often command higher prices in the market due to their perceived quality and sustainability. The Rodale Institute's research highlights the potential for biodynamic farms to achieve premium pricing and improved profitability.

2. Economic Resilience

The institute's studies suggest that biodynamic farming can enhance economic resilience by reducing reliance on external inputs and improving farm sustainability.

- **Self-Sufficiency:** Biodynamic practices promote self-sufficiency through the use of on-farm resources and closed nutrient cycles. This can reduce the impact of price fluctuations and supply chain disruptions.
- **Resilience to Climate Change:** Biodynamic farms may be better equipped to adapt to climate change due to their emphasis on soil health and ecological balance. Improved soil structure and organic matter content can enhance water retention and reduce vulnerability to extreme weather events.

Market Trends and Impacts

1. Consumer Demand

The Rodale Institute's research indicates that consumer demand for biodynamic and sustainably produced products is growing. Consumers are increasingly interested in products that are perceived as healthier, more environmentally friendly, and ethically produced.

- **Market Growth:** The demand for biodynamic products is part of a broader trend towards organic and sustainable agriculture. The Rodale Institute's work helps to validate and support this demand by providing scientific evidence of the benefits of biodynamic farming.
- **Consumer Education:** The institute's educational programs and outreach efforts contribute to increasing consumer awareness and understanding of biodynamic practices, which can drive market growth and support for biodynamic products.

2. Retail and Distribution

The Rodale Institute's research also explores the challenges and opportunities associated with marketing and distributing biodynamic products.

- **Niche Markets:** Biodynamic products are often marketed through specialty retailers, farmer's markets, and direct-to-consumer channels. These markets offer opportunities for premium pricing and direct engagement with consumers.
- **Mainstream Integration:** As awareness of biodynamic practices grows, there is potential for biodynamic products to enter mainstream retail channels. The Rodale Institute's research supports efforts to integrate biodynamic products into larger retail networks and distribution systems.

6.2 Case Study 2: Demeter Biodynamic Farm, Australia

Demeter Biodynamic Farm in Australia is a prime example of a successful biodynamic operation. The farm's experience in managing production costs, achieving financial sustainability, and navigating market challenges provides valuable lessons for other biodynamic farmers. The farm's focus on premium pricing and niche markets demonstrates the potential for profitability in biodynamic

farming.

6.3 Case Study 3: Sekem, Egypt

Sekem, founded by Ibrahim Abouleish, is a pioneering biodynamic farm in Egypt that transformed arid land into fertile farmland using biodynamic practices. The farm's success in creating a thriving ecosystem and generating economic value highlights the impact of biodynamic farming in challenging environments and its role in sustainable development.

Conclusion

Biodynamic farming offers a unique approach to agriculture that integrates ecological principles with holistic practices. While the environmental and ethical benefits are well-established, understanding the economic and market considerations is essential for assessing the viability and impact of biodynamic farming. This article has explored the production costs, financial sustainability, market demand, and pricing strategies associated with biodynamic farming. By analysing case studies and market trends, it provides insights into the opportunities and challenges faced by biodynamic farmers. The future of biodynamic farming will depend on continued innovation, consumer education, and effective market strategies to ensure its growth and sustainability within the broader agricultural industry.

References

Koepf, H. H. (2005). The Biodynamic Farm: Agriculture in Service of the Earth and Humanity. SteinerBooks.

Muhie, S. H. (2023). Concepts, principles, and application of biodynamic farming: a review. Circular Economy and Sustainability, 3(1), 291-304.

Weintraub, I. (2002). Farming and Farming Systems. In Using the Agricultural, Environmental, and Food Literature (pp. 172-216). CRC Press.

Wierzbicki, M. (2012). Biodynamic farming and nutrition and the conscious evolution of consumer society.

10

Challenges and Criticisms Associated with Biodynamic Farming

Abstract

Biodynamic farming, a holistic approach to agriculture pioneered by Rudolf Steiner, integrates ecological principles, cosmic rhythms, and spiritual practices into farming methods. While it offers numerous environmental and ethical benefits, biodynamic farming also faces various challenges and criticisms. This article examines the key challenges and criticisms associated with biodynamic farming, including scientific skepticism, economic feasibility, practical implementation issues, and market acceptance. Through a detailed analysis of these aspects, the article aims to provide a comprehensive understanding of the limitations and contentious aspects of biodynamic farming, offering insights into potential areas for improvement and future research.

Introduction

Biodynamic farming, developed by Rudolf Steiner in the early 20th century, represents a unique and holistic approach to agriculture. It emphasizes the integration of ecological balance, soil health, cosmic rhythms, and spiritual practices into farming methods. Despite its growing popularity and recognition for its environmental and ethical benefits, biodynamic farming is not without its challenges and criticisms. This article explores the key challenges and criticisms associated with biodynamic farming, examining issues related to scientific validation, economic viability, practical implementation, and market acceptance.

1. Scientific Skepticism

One of the primary criticisms of biodynamic farming is the perceived lack of empirical evidence supporting its effectiveness. Critics argue that many of the principles and practices of biodynamic farming are based on spiritual and esoteric beliefs rather than scientific evidence.

DOI: 10.1201/9781042015597-10

1.1 Cosmic Rhythms and Preparations

Biodynamic farming, founded by Rudolf Steiner, incorporates a unique set of practices and principles that blend agricultural science with spiritual and cosmic elements. Among these practices, cosmic rhythms and biodynamic preparations are central to the methodology. This section elaborates on these concepts, their Cosmic rhythms refer to the belief in the influence of celestial bodies, such as the moon, planets, and stars, on agricultural processes. In biodynamic farming, these cosmic influences are thought to affect plant growth, soil health, and overall farm vitality.

1.1.1 Planetary Influences

- **Planetary Movements:** Biodynamic practices also consider the positions of planets and their alignment with agricultural activities. Specific planetary configurations are believed to influence different aspects of plant growth and soil health.
- **Astrological Significance:** Some biodynamic practitioners use astrological charts to determine the best times for planting and harvesting, based on the alignment of celestial bodies.

1.1.2 Challenges

1. **Lack of Empirical Evidence**
    - **Scientific Validation:** The scientific community has generally been skeptical about the influence of cosmic rhythms on plant growth. Empirical evidence supporting the impact of lunar phases and planetary movements on agricultural outcomes is limited. Studies often show mixed results, with many failing to demonstrate a clear and consistent effect.
    - **Mechanisms of Influence:** The mechanisms by which cosmic rhythms might affect plants are not well-understood or scientifically validated. Critics argue that the observed benefits of following lunar cycles may be attributed to other factors, such as improved planning and management.
2. **Practical Implementation Issues**
    - **Complexity:** Incorporating cosmic rhythms into farming practices adds a layer of complexity. Farmers must synchronize their activities with lunar and planetary cycles, which can be challenging to manage alongside other farming tasks.

- **Inconsistencies:** The influence of cosmic rhythms may vary depending on geographical location and local environmental conditions, leading to inconsistencies in practice and outcomes.

1.2 Biodynamic Preparations : Concept and Usage

Biodynamic preparations are natural substances used in biodynamic farming to enhance soil health, plant growth, and overall farm vitality. These preparations are made from various organic materials and are applied to the soil, compost, or plants according to specific methods.

1.2.1 Challenges

1. **Effectiveness and Validation**
   - **Scientific Research:** The effectiveness of biodynamic preparations is a topic of debate. Some studies suggest that these preparations may improve soil health and plant growth, while others find no significant effects. The lack of consistent and rigorous scientific research makes it difficult to validate the claims associated with these preparations.
   - **Mechanisms of Action:** The mechanisms by which biodynamic preparations are thought to work are not always well-understood. Critics argue that the observed benefits may be due to other factors, such as improved compost quality or enhanced soil management practices.
2. **Practical Considerations**
   - **Preparation Methods:** The preparation of biodynamic substances involves specific rituals and methods, such as burying materials in cow horns. These practices may be viewed as unscientific or impractical by those who prioritize empirical evidence and modern technology.
   - **Labour and Resource Intensity:** The production and application of biodynamic preparations require additional labor and resources. This can be a burden for farmers, particularly those with limited resources or those transitioning from conventional farming practices.
3. **Scientific Research and Validation**
   - **Research Limitations:** There is a need for more rigorous and large-scale scientific studies to validate the claims of biodynamic farming. Some research suggests benefits such as improved soil health and biodiversity, but more extensive and peer-reviewed studies are needed to confirm these findings.

- **Comparative Studies:** Comparative studies between biodynamic and conventional or organic farming systems often show mixed results. While some studies highlight the advantages of biodynamic practices, others find no significant differences in yield or environmental impact.

2. Perceived Unscientific Aspects

2.1 Spiritual and Esoteric Beliefs

- **Foundational Beliefs:** Biodynamic farming is rooted in Steiner's spiritual and esoteric beliefs, which include the idea of a farm as a living organism interconnected with cosmic forces. Critics argue that these beliefs lack scientific basis and may deter scientifically-minded individuals from adopting biodynamic practices.
- **Subjectivity:** The subjective nature of some biodynamic practices, such as the use of specific preparations and rituals, can lead to skepticism among those who prioritize empirical and quantifiable evidence.

3. Economic Feasibility

3.1 High Production Costs

1. **Certification and Input Costs**
    - **Certification Fees:** Obtaining and maintaining biodynamic certification can be costly for farmers. The certification process involves initial fees, annual inspections, and compliance with specific standards, which can be a financial burden, especially for small-scale farmers.
    - **Input Costs:** Biodynamic farming often requires the use of specialized inputs and preparations, such as compost preparations and organic seeds. These inputs can be more expensive than conventional alternatives, leading to higher production costs.
2. **Labor and Resource Requirements**
    - Manual Labor: Biodynamic farming typically involves more manual labour compared to conventional farming. Tasks such as composting, manual weeding, and animal care can increase labour costs and require additional workforce.
    - Resource Allocation: Biodynamic practices may necessitate the allocation of resources to maintain soil health and ecological balance, which can affect overall farm profitability.

3.2 Market Challenges

1. **Premium Pricing**
    - **Consumer Willingness:** While biodynamic products often command premium prices, the willingness of consumers to pay higher prices can vary. Economic downturns and price sensitivity can impact the marketability and profitability of biodynamic products.
    - **Market Segmentation:** Biodynamic products are often marketed through specialty retailers and niche markets. Expanding into mainstream markets requires effective marketing strategies and may face competition from organic and conventional products.
2. **Economic Resilience**
    - Market Fluctuations: Biodynamic farms may face challenges in achieving economic resilience due to fluctuating market demand and input costs. The reliance on premium pricing and niche markets can make biodynamic farms vulnerable to economic shifts.

3.3 Practical Implementation Issues

1. **Technical Knowledge and Skills**
    - Specialized Knowledge: Biodynamic farming requires specialized knowledge and skills related to soil management, plant growth, and the use of biodynamic preparations. Farmers may need extensive training and education to effectively implement biodynamic practices.
    - **Learning Curve:** The complexity of biodynamic methods can present a steep learning curve for farmers transitioning from conventional or organic farming systems. This can lead to challenges in achieving desired outcomes and maintaining farm productivity.
2. **Operational Challenges**
    - Resource Management: Managing resources such as compost, animal manure, and biodynamic preparations can be challenging. Ensuring the proper preparation and application of these resources requires careful attention and adherence to specific practices.
    - Record Keeping: Biodynamic farming involves meticulous record-keeping to monitor cosmic cycles, preparation use, and other aspects. Maintaining accurate records can be time-consuming and requires additional effort from farmers.

3.4 Integration with Conventional Practices

1. **Transition Period**
   - Adapting Practices: Transitioning to biodynamic farming from conventional practices can be a complex process. Farmers may face challenges in adapting their methods, managing soil health, and achieving desired results during the transition period.
   - Yield and Productivity: During the transition phase, farms may experience temporary declines in yield and productivity as they adapt to new practices. This can impact short-term economic viability and farm performance.
2. **Compatibility with Modern Agriculture**
   - Technological Integration: Integrating biodynamic practices with modern agricultural technologies can be challenging. Some biodynamic principles may conflict with conventional technological approaches, requiring careful consideration of compatibility.
   - Policy and Regulation: Biodynamic farming may face regulatory and policy challenges related to certification, labelling, and market access. Navigating these regulatory requirements can be complex and may impact farm operations.

3.5 Market Acceptance and Consumer Perception

3.5.1 Consumer Awareness

1. **Understanding Biodynamic Principles**
   - Education and Awareness: Many consumers may lack a clear understanding of biodynamic farming principles and practices. Educating consumers about the benefits and unique aspects of biodynamic products is essential for increasing market acceptance.
   - Perceived Value: The perceived value of biodynamic products can be influenced by consumer awareness and understanding. Effective communication and marketing strategies are necessary to highlight the benefits of biodynamic farming.
2. **Market Differentiation**
   - Product Differentiation: Biodynamic products are often positioned as premium offerings in the market. Differentiating these products from organic and conventional alternatives requires effective branding and messaging.
   - Market Competition: Biodynamic products face competition from organic and conventional products, which may offer similar benefits at

lower prices. Differentiating biodynamic products and demonstrating their unique value is crucial for market success.

3.5.2 Certification Criticisms

1. **Certification Standards**
    - **Consistency and Transparency:** Critics argue that certification standards for biodynamic farming may lack consistency and transparency. Ensuring that certification bodies adhere to rigorous and transparent standards is important for maintaining credibility and trust.
    - **Certification Process:** The certification process can be perceived as complex and bureaucratic, potentially deterring some farmers from pursuing biodynamic certification. Streamlining the process and reducing barriers to certification may enhance accessibility.

2. Certification Costs

Certification is a pivotal component of biodynamic farming, distinguishing it from other agricultural practices and ensuring adherence to specific standards. However, the costs associated with certification can present significant challenges for farmers. These costs encompass various aspects of the certification process, each contributing to the overall financial burden experienced by those transitioning to or maintaining biodynamic practices.

Initial Certification Fees : The initial certification fees represent one of the primary costs associated with biodynamic farming. These fees are charged by certification bodies to evaluate and verify that a farm meets the standards required for biodynamic certification. The initial certification process typically involves:

1. **Application Fees:** Farmers must pay application fees when applying for biodynamic certification. These fees cover the administrative costs associated with processing the application and conducting preliminary assessments.
2. **Inspection Fees:** An essential part of the certification process is the on-site inspection conducted by certification bodies. Inspectors evaluate farm practices, review records, and assess compliance with biodynamic standards. The cost of these inspections can vary based on the size of the farm and the complexity of its operations.
3. **Documentation and Record-Keeping:** Certification requires extensive documentation and record-keeping to demonstrate compliance with biodynamic principles. This can involve additional costs related to

maintaining accurate records and preparing necessary documentation for inspection.

3. Ongoing Certification Costs

Once certified, farms are subject to ongoing costs associated with maintaining their biodynamic status. These ongoing costs include:

1. **Annual Fees:** Biodynamic certification bodies charge annual fees to cover the costs of routine inspections, monitoring, and certification renewal. These fees ensure that farms continue to meet the standards and remain in compliance with biodynamic practices.
2. **Renewal Costs:** Certification is not a one-time process; farms must undergo periodic renewals to maintain their certification status. Renewal costs can be substantial, as they often include re-inspections, updated documentation, and administrative expenses.
3. **Compliance Costs:** Maintaining certification requires ongoing adherence to biodynamic standards, which may involve additional costs. These can include investments in specific inputs, such as biodynamic preparations and compost, as well as adjustments to farming practices to remain compliant.

3.5.2 Financial Impact on Farmers

The financial impact of certification costs can be particularly challenging for small-scale and emerging biodynamic farms. The expenses associated with obtaining and maintaining certification can strain the budgets of farms that are already operating on tight margins. Key financial considerations include:

1. **Economic Feasibility:** The costs of certification can be a significant barrier for new or small-scale farmers considering biodynamic practices. The initial investment, combined with ongoing fees, may outweigh the potential benefits for some farmers, impacting their decision to pursue certification.
2. **Return on Investment:** While biodynamic certification can enhance marketability and potentially command premium prices for products, the financial return on investment is not always guaranteed. Farmers must carefully weigh the costs against the potential benefits, including market access, consumer demand, and product differentiation.
3. **Access to Support:** Financial support and subsidies for certification costs may be available through various programs, grants, and organizations dedicated to promoting sustainable agriculture. Accessing these

resources can help alleviate some of the financial pressures associated with certification.

3.5.3 Mitigating Certification Costs

Farmers can explore several strategies to mitigate the costs associated with biodynamic certification:

1. **Gradual Transition:** Implementing biodynamic practices gradually and seeking partial certification for specific areas or crops can help manage initial costs. This approach allows farmers to spread out expenses over time and adapt to biodynamic methods at a manageable pace.
2. **Cooperatives and Shared Resources:** Joining biodynamic farming cooperatives or networks can provide access to shared resources and collective certification processes. These arrangements can reduce individual costs and offer support through shared expertise and knowledge.
3. **Financial Planning:** Developing a comprehensive financial plan that accounts for certification costs and ongoing expenses is essential for managing the financial impact. Farmers should budget for certification fees, explore funding options, and assess the long-term financial implications of certification.

Conclusion

Biodynamic farming presents a unique and holistic approach to agriculture, emphasizing ecological balance, soil health, and cosmic rhythms. While it offers significant environmental and ethical benefits, it also faces various challenges and criticisms related to scientific skepticism, economic feasibility, practical implementation, and market acceptance. Addressing these challenges requires continued research, education, and dialogue to enhance the understanding and adoption of biodynamic practices. By addressing criticisms and improving the practical aspects of biodynamic farming, the agricultural community can work towards a more sustainable and resilient future.

References

Kristiansen, P., Taji, A., &Reganold, J. (2006). Organic agriculture: opportunities and challenges. Organic Agriculture: A Global Perspective. CSIRO Publishing, Australia, 421-441.

Oberč, B. P., & Arroyo Schnell, A. (2020). Approaches to sustainable agriculture. Exploring the pathways, 486.

Rigby, D., & Caceres, D. (1997). The sustainability of agricultural systems.

Rigolot, C. (2023). Biodynamic farming research and transdisciplinary knowledge co-production: Exploring the synergies. GAIA-Ecological Perspectives for Science and Society, 32(4), 353-358.

11

Future of Biodynamic Farming: Emerging Trends and Technologies

Abstract

Biodynamic farming, an agricultural approach founded on Rudolf Steiner's principles, has evolved significantly since its inception. As the global agricultural landscape continues to shift towards sustainability, the future of biodynamic farming is increasingly influenced by emerging trends and technologies. This article explores the future trajectory of biodynamic farming, focusing on key trends such as the integration of digital technologies, advancements in soil health management, and the growing emphasis on climate resilience. Additionally, the article examines how biodynamic practices are adapting to meet contemporary challenges and opportunities. By analyzing these developments, the article aims to provide a comprehensive overview of the future potential of biodynamic farming and its role in shaping sustainable agriculture.

Introduction

Biodynamic farming, established by Rudolf Steiner in the early 20th century, is characterized by its holistic approach to agriculture, which integrates ecological balance, cosmic rhythms, and spiritual principles. As agriculture faces increasing pressures from environmental degradation, climate change, and the need for sustainable practices, biodynamic farming is evolving to address these challenges. The future of biodynamic farming is marked by emerging trends and technologies that enhance its principles while adapting to contemporary agricultural needs. This article explores these developments and assesses their potential impact on the future of biodynamic farming.

I. Emerging Trends in Biodynamic Farming

1.1 Integration of Digital Technologies

The integration of digital technologies represents a transformative opportunity for biodynamic farming, aligning traditional agricultural practices with modern innovations. By incorporating tools such as sensors, data analytics, and digital platforms, biodynamic farmers can enhance their practices, improve efficiency,

DOI: 10.1201/9781042015597-11

and achieve more precise outcomes. This section explores the various aspects of digital technology integration in biodynamic farming, including its benefits, challenges, and potential applications.

1.1.1 Precision Agriculture: Data-Driven Decision Making

1. **Sensors and Monitoring Systems**
    - **Soil Sensors:** Soil sensors measure key parameters such as moisture levels, temperature, pH, and nutrient content. These real-time data points enable biodynamic farmers to make informed decisions about soil management and the application of biodynamic preparations. For example, sensors can help determine the optimal timing for compost application or the need for specific soil amendments.
    - **Crop Health Monitoring:** Remote sensing technologies, including drones and satellite imagery, provide valuable insights into crop health and growth. By analyzing images and data collected through these technologies, farmers can detect issues such as nutrient deficiencies, pest infestations, or diseases early, allowing for timely interventions.
2. **Precision Application**
    - **Targeted Inputs:** Precision agriculture technologies enable farmers to apply biodynamic preparations and other inputs with greater accuracy. For instance, variable-rate technology allows for the targeted application of compost or biodynamic preparations based on specific soil and crop conditions. This reduces waste and maximizes the effectiveness of inputs.
    - **Automated Systems:** Automated systems, including drones and robotic equipment, can perform tasks such as planting, irrigation, and fertilization with high precision. Integration with biodynamic principles allows these systems to align with lunar cycles and other biodynamic practices, optimizing resource use and improving overall farm management.

1.1.2 Data Integration and Management

1. **Farm Management Software**
    - **Planning and Scheduling:** Farm management software helps biodynamic farmers plan and schedule their activities, including planting, harvesting, and the application of biodynamic preparations. These tools often include features such as calendar integration, task management, and record-keeping, facilitating adherence to biodynamic principles.

- **Performance Tracking:** Software platforms enable farmers to track performance metrics such as crop yields, soil health, and input usage. By analyzing historical data and trends, farmers can identify patterns, optimize practices, and make data-driven decisions to enhance farm productivity.

2. **Data Analytics**
   - **Predictive Analytics:** Advanced data analytics tools use historical and real-time data to predict future outcomes and optimize decision-making. For example, predictive models can forecast weather patterns, pest outbreaks, or soil conditions, helping farmers plan their activities more effectively and mitigate potential risks.
   - **Performance Evaluation:** Data analytics allows farmers to evaluate the impact of biodynamic practices on farm performance. By analyzing data on soil health, crop yields, and other indicators, farmers can assess the effectiveness of biodynamic methods and make informed adjustments to their practices.

1.1.3 Digital Platforms and Tools : Online Communities and Knowledge Sharing

1. **Digital Networks**
   - **Collaboration and Support:** Online communities and networks dedicated to biodynamic farming provide platforms for collaboration, knowledge sharing, and support. Farmers can connect with peers, experts, and organizations to exchange experiences, seek advice, and stay informed about the latest developments in biodynamic practices.
   - **Educational Resources:** Digital platforms offer access to educational resources such as webinars, articles, and training materials. These resources can help farmers learn about new technologies, techniques, and research findings related to biodynamic farming.
2. **Resource Access**
   - **Market Access:** Online platforms facilitate access to markets for biodynamic products, connecting farmers with consumers, retailers, and distributors. This can enhance market visibility and opportunities for selling biodynamic products, contributing to the economic viability of biodynamic farming.
   - **Supplier Networks:** Digital tools can help farmers identify and connect with suppliers of biodynamic inputs, including compost materials, preparations, and equipment. This streamlines the procurement process and ensures access to high-quality resources.

1.1.4 Block chain and Traceability

1. **Transparency and Authentication**
    - **Supply Chain Tracking:** Blockchain technology enables transparent tracking of biodynamic products throughout the supply chain. This enhances traceability, allowing consumers and stakeholders to verify the authenticity of biodynamic claims and ensure compliance with certification standards.
    - **Certification Verification:** Blockchain can streamline the verification process for biodynamic certification, providing a secure and tamper-proof record of compliance. This can reduce administrative burdens and increase trust in biodynamic practices.
2. **Data Security**
    - **Secure Transactions:** Blockchain technology provides a secure platform for recording and managing transactions, including financial transactions and certification records. This enhances data security and reduces the risk of fraud or data manipulation.

1.1.5 Challenges and Considerations

a. Cost and Accessibility

1. **Initial Investment**
    - **Technology Costs:** The initial investment in digital technologies, such as sensors, drones, and farm management software, can be substantial. This may pose a barrier for small-scale or emerging biodynamic farms with limited financial resources.
    - **Ongoing Expenses:** Maintenance, software subscriptions, and data management costs can add to the financial burden of integrating digital technologies. Farmers must weigh these costs against the potential benefits and returns on investment.
2. **Technical Expertise**
    - **Skill Requirements:** Integrating digital technologies requires technical expertise and training. Farmers may need to invest in learning new skills or hiring experts to effectively use and manage these technologies.
    - **Support and Training:** Access to support and training resources is crucial for successful technology integration. Farmers should seek out educational opportunities and support networks to ensure they can effectively implement and utilize digital tools.

b. Data Management and Privacy

1. Data Privacy

- **Protecting Sensitive Information:** Managing and securing data related to farm operations, including soil health, crop performance, and financial information, is essential. Farmers must be aware of data privacy issues and take measures to protect sensitive information.
- **Compliance:** Compliance with data protection regulations and standards is important for safeguarding data and maintaining trust with stakeholders. Farmers should be informed about relevant regulations and best practices for data management.

2. Data Integration

- **Interoperability:** Integrating data from multiple sources and technologies can be challenging. Ensuring interoperability between different systems and platforms is crucial for effective data management and decision-making.
- **Data Quality:** The accuracy and reliability of data are critical for making informed decisions. Farmers must ensure that data collected through digital technologies is accurate and representative of farm conditions.

c. Advancements in Soil Health Management

1. Soil Microbiome Research

- **Understanding Soil Microbes:** Advances in soil microbiome research are enhancing our understanding of the complex interactions between soil microbes, plants, and biodynamic preparations. Research into soil microbiomes can provide insights into how biodynamic practices influence soil health and plant growth at the microbial level.
- **Soil Health Indicators:** New methods for assessing soil health, including microbiome analysis and soil chemistry tests, can help biodynamic farmers monitor and improve soil conditions. These indicators can inform the application of biodynamic preparations and support the development of more effective soil management practices.

2. Innovative Composting Techniques

- **Enhanced Composting Methods:** Innovations in composting technology, such as aerobic and vermicomposting systems, can improve the quality and efficacy of biodynamic compost preparations. These techniques can help optimize nutrient content, microbial activity, and soil fertility.

- **Integration with Biodynamic Practices:** Combining traditional biodynamic composting methods with modern composting technologies can enhance the benefits of biodynamic compost preparations. This integration can improve soil health and support sustainable farming practices.

d. Climate Resilience and Adaptation

1. Climate-Smart Practices

- **Drought and Heat Resistance:** Biodynamic farming practices can be adapted to enhance resilience to climate-related challenges such as drought and heat. Techniques such as improved soil moisture retention, cover cropping, and mulching can help mitigate the impacts of climate change on crop yields and soil health.
- **Carbon Sequestration:** Biodynamic practices that focus on soil health and organic matter contribute to carbon sequestration, helping to mitigate climate change. Enhanced understanding of soil-carbon dynamics can inform strategies for maximizing carbon storage in biodynamic systems.

2. Biodiversity and Ecosystem Services:

- **Promoting Biodiversity:** Biodynamic farming emphasizes biodiversity and ecosystem services, which can enhance resilience to climate change. Strategies such as habitat conservation, crop diversification, and the integration of beneficial organism's support ecosystem stability and resilience.
- **Ecosystem-Based Adaptation:** Adapting biodynamic practices to support ecosystem-based adaptation can improve the overall resilience of farming systems. This includes strategies to enhance soil health, water management, and habitat quality.

e. Technologies Shaping the Future of Biodynamic Farming

1. Soil and Crop Sensors

- **Real-Time Monitoring:** Soil and crop sensors provide real-time data on soil moisture, temperature, nutrient levels, and crop health. These sensors can help biodynamic farmers make informed decisions about the application of biodynamic preparations and resource management.
- **Precision Application:** Sensors enable precision application of biodynamic preparations and inputs, optimizing their effectiveness and reducing waste. This aligns with the biodynamic principle of maximizing resource efficiency.

2. **Internet of Things (IoT)**
   - **Connected Farm Systems:** IoT technologies facilitate the integration of various farm systems, including sensors, irrigation systems, and farm management software. This connectivity allows for more efficient management of biodynamic practices and resources.
   - **Data Integration:** IoT platforms can integrate data from multiple sources, providing a comprehensive view of farm conditions and supporting data-driven decision-making in biodynamic farming.

f. Genetic and Breeding Innovations

1. **Plant Breeding**
   - **Resilient Varieties:** Advances in plant breeding can produce crop varieties that are more resilient to environmental stresses and pests. These innovations can complement biodynamic practices by enhancing crop performance and reducing the need for external inputs.
   - **Traditional and Modern Approaches:** Combining traditional breeding techniques with modern genetic tools can create new crop varieties that align with biodynamic principles and contribute to sustainable farming practices.
2. **Soil Microbe Selection**
   - **Microbial Inoculants:** Research into soil microbes and microbial inoculants can enhance soil health and fertility in biodynamic systems. Selecting and applying beneficial microbes can improve soil structure, nutrient availability, and plant growth.
   - **Customized Solutions:** Tailoring microbial inoculants to specific soil conditions and crop needs can optimize the benefits of biodynamic practices and support farm productivity.

g. Integration of Renewable Energy

1. **Renewable Energy Sources**
   - **Solar and Wind Power:** Incorporating renewable energy sources such as solar and wind power into biodynamic farming operations can reduce reliance on fossil fuels and lower energy costs. Renewable energy can power farm machinery, irrigation systems, and processing facilities.
   - **Energy Efficiency:** Implementing energy-efficient practices and technologies can complement biodynamic principles by reducing

the environmental impact of farming operations and supporting sustainable resource use.

2. **On-Farm Energy Solutions**
   - **Biogas Production:** On-farm biogas production from organic waste and manure can provide a renewable energy source and enhance nutrient recycling in biodynamic systems. Biogas can be used for energy and heat, reducing the need for external inputs.
   - **Energy Storage:** Advances in energy storage technologies, such as batteries and thermal storage, can support the integration of renewable energy into biodynamic farming operations. Effective energy storage solutions can enhance farm resilience and sustainability.

2. Adapting Biodynamic Practices for the Future

Biodynamic farming, rooted in principles established by Rudolf Steiner, is evolving to address contemporary agricultural challenges while remaining true to its foundational beliefs. Adapting biodynamic practices for the future involves integrating modern innovations, embracing holistic approaches, and responding to environmental and market demands. This section explores the key strategies for adapting biodynamic practices, including incorporating modern technologies, addressing climate challenges, and enhancing collaboration and knowledge sharing.

2.1 Holistic and Integrated Approaches : Systems Thinking

2.1.1 Farm as an Ecosystem

- **Interconnected Management:** Viewing the farm as an interconnected ecosystem is central to adapting biodynamic practices. This approach emphasizes the integration of soil health, crop management, livestock, and broader environmental factors. By managing these elements as a cohesive system, farmers can enhance overall farm sustainability and resilience.
- **Synergies and Trade-Offs:** Understanding the synergies and trade-offs between different biodynamic practices allows for more effective and efficient management. For example, integrating livestock for manure production can benefit soil health, while careful management is needed to balance nutrient application and avoid overuse.

2.1.2 Ecosystem Services

- **Enhancing Services:** Biodynamic farming can benefit from a focus on ecosystem services, such as pollination, soil fertility, and water management. Practices that support these services, such as planting cover crops and creating wildlife habitats, contribute to farm resilience and productivity.

- **Adaptive Strategies:** Adapting practices to enhance ecosystem services involves implementing strategies that address specific environmental conditions and challenges. This may include adjusting crop rotations, incorporating agroforestry, or improving water conservation techniques.

2.2 Collaborative Research and Innovation

2.2.1 Partnerships and Networks

- **Research Collaboration:** Engaging in collaborative research partnerships with academic institutions, agricultural organizations, and industry stakeholders can drive advancements in biodynamic farming. These partnerships can facilitate the development and adoption of new technologies, practices, and insights.
- **Farmer Networks:** Participating in farmer networks and cooperatives allows for the exchange of knowledge and experiences. Collaborative efforts within these networks can lead to shared solutions and innovations that benefit the broader biodynamic community.

2.2.2 **Knowledge Sharing**

2.2.3 Education and Training

Ongoing education and training are essential for adapting biodynamic practices. Access to workshops, webinars, and other educational resources helps farmers stay informed about new developments and techniques.

2.3.4 Best Practices

Sharing best practices and success stories within the biodynamic community fosters a culture of continuous improvement. Farmers can learn from each other's experiences and adapt successful strategies to their own operations.

2.3 Addressing Climate Challenges: Climate-Smart Practices

2.3.1 Resilience to Environmental Stresses

- **Drought and Heat Adaptation:** Biodynamic practices can be adapted to improve resilience to climate-related stresses such as drought and heat. Techniques like improved soil moisture retention, use of drought-resistant crop varieties, and efficient irrigation systems help mitigate the impacts of changing climate conditions.
- **Frost and Extreme Weather:** Strategies to manage frost and extreme weather events include implementing frost protection measures, such as

windbreaks and frost-resistant crops, and developing contingency plans for weather-related disruptions.

2.3.2 Carbon Sequestration

- **Soil Carbon Storage:** Biodynamic practices that focus on soil health, such as composting and cover cropping, contribute to carbon sequestration. Enhancing soil carbon storage helps mitigate climate change and improves soil fertility.
- **Measurement and Verification:** Measuring and verifying carbon sequestration benefits can provide valuable data for assessing the impact of biodynamic practices on climate mitigation. Farmers can use this information to support sustainability claims and access carbon markets.

2.4 Biodiversity and Ecosystem Adaptation

2.4.1 Promoting Biodiversity

- **Habitat Conservation:** Biodynamic farming practices that support biodiversity, such as creating wildlife habitats and planting diverse crops, enhance ecosystem resilience. Biodiversity conservation helps maintain ecological balance and supports farm productivity.
- **Pollinator Support:** Implementing practices that support pollinators, such as planting flowering plants and avoiding harmful pesticides, contributes to healthy ecosystems and improves crop yields.

2.4.2 Ecosystem-Based Adaptation:

- **Integrated Approaches:** Adapting biodynamic practices to support ecosystem-based adaptation involves integrating practices that enhance overall ecosystem health. This includes strategies such as agroforestry, wetland restoration, and sustainable land management.
- **Local Adaptation:** Tailoring practices to local environmental conditions and challenges ensures that adaptations are effective and relevant. Local adaptation strategies may involve adjusting crop selections, soil management practices, and water conservation techniques.

2.5 Embracing Technological Innovations: Digital Tools and Technologies

2.5.1 Advanced Monitoring and Analytics

- **Real-Time Data:** Leveraging advanced monitoring tools, such as sensors and drones, provides real-time data on soil conditions, crop health, and environmental factors. This data enables farmers to make informed decisions and optimize biodynamic practices.

- **Data Analytics:** Utilizing data analytics tools to interpret and analyze data helps farmers identify trends, optimize practices, and address challenges. Predictive models and performance evaluations support continuous improvement in biodynamic farming.

2.5.2 Automation and Precision

- **Automated Systems:** Integrating automation technologies for tasks such as irrigation, planting, and nutrient application enhances precision and efficiency. Automated systems align with biodynamic principles by reducing resource waste and optimizing input use.
- **Precision Agriculture:** Applying precision agriculture techniques, including variable-rate technology and targeted inputs, allows for more accurate application of biodynamic preparations and inputs.

2.6 Innovative Practices

2.6.1 Genetic and Breeding Innovations

- **Resilient Varieties:** Advances in plant breeding can produce crop varieties that are more resilient to environmental stresses and pests. These innovations complement biodynamic practices by enhancing crop performance and reducing the need for external inputs.
- **Microbial Inoculants:** Research into beneficial soil microbes and microbial inoculants supports soil health and fertility in biodynamic systems. Customized microbial solutions can optimize soil conditions and support plant growth.

1.1.2 Renewable Energy Integration

- **Energy Efficiency:** Incorporating renewable energy sources, such as solar and wind power, into biodynamic farming operations reduces reliance on fossil fuels and supports sustainable resource use. Energy-efficient technologies further enhance farm sustainability.
- **On-Farm Solutions:** Implementing on-farm renewable energy solutions, such as biogas production and energy storage, contributes to farm resilience and reduces environmental impact.

Conclusion

The future of biodynamic farming is shaped by emerging trends and technologies that enhance its principles while addressing contemporary agricultural challenges. Integration of digital technologies, advancements in soil health management, and a focus on climate resilience are driving the evolution of biodynamic practices. By embracing innovations such as sensor technologies, genetic advancements, and renewable energy solutions, biodynamic farming can continue to thrive and contribute to sustainable agriculture.

As biodynamic farming adapts to new developments, it is essential for practitioners to remain informed and engaged with emerging trends and technologies. By leveraging these advancements and maintaining a holistic approach, biodynamic farming can play a pivotal role in shaping the future of agriculture and promoting environmental stewardship.

References

Muhie, S. H. (2022). Novel approaches and practices to sustainable agriculture. Journal of Agriculture and Food Research, 10, 100446.

Paull, J., & Hennig, B. (2020). A world map of biodynamic agriculture. Agricultural and Biological Sciences Journal, 6(2), 114-119.

Sani, S., Abdulkadir, A. I., Abdullahi, N., Abubakar, K. M., Abdul, A., & Umar, A. L. Modern Approaches to Sustainable Agriculture.

Sharma, R., Aravind, T., & Sharma, R. (2019). Sustainable agriculture: Trends and opportunities for 21st Century. Journal of Applied and Natural Science, 11(3), 666-672.

12

Profiles of Notable Biodynamic Farms and Startup Needs to Biodynamic Farm

Abstract

Biodynamic farming, with its holistic approach to agriculture, has seen significant growth and interest in recent years. This article explores profiles of notable biodynamic farms that exemplify the principles and practices of biodynamic agriculture. It highlights the successes and innovations of these farms, offering insights into their operations, challenges, and contributions to sustainable agriculture. Additionally, the article addresses the startup needs for establishing a biodynamic farm, covering essential considerations such as land selection, certification, initial investments, and resources. By examining these profiles and startup requirements, the article provides a comprehensive overview of the current state and future potential of biodynamic farming.

Introduction

Biodynamic farming, an approach founded by Rudolf Steiner, emphasizes a holistic and ecological approach to agriculture. This farming method integrates soil health, cosmic rhythms, and spiritual principles to achieve sustainable and resilient farming systems. As interest in biodynamic farming grows, notable farms around the world exemplify the success and innovation possible within this framework. This article profiles several exemplary biodynamic farms and outlines the essential startup needs for those interested in beginning a biodynamic farming venture.

1. Profiles of Notable Biodynamic Farms

1.1 Demeter Farm, Germany

Overview: Demeter Farm, located in the heart of Germany, is a leading example of biodynamic farming. Established in the early 1920s, it has been a pioneer in implementing Rudolf Steiner's principles. The farm operates on a diverse range of crops, including vegetables, grains, and fruits, and integrates livestock to enhance soil fertility and ecosystem balance.

DOI: 10.1201/9781042015597-12

Key Features

- **Comprehensive Biodynamic Practices:** Demeter Farm adheres to all aspects of biodynamic farming, including the use of biodynamic preparations, lunar planting cycles, and composting.
- **Education and Outreach:** The farm offers educational programs and workshops for farmers and the public, promoting biodynamic principles and sustainable farming practices.
- **Research and Innovation:** Demeter Farm conducts ongoing research into soil health, crop performance, and the impact of biodynamic methods, contributing to the broader understanding of biodynamic farming.

Successes

- **Soil Health:** The farm has achieved remarkable improvements in soil health and fertility through the use of biodynamic composts and preparations.
- **Community Engagement:** Demeter Farm has successfully engaged with local communities through educational initiatives and farmer networks.

Challenges

- **Certification Requirements:** Maintaining biodynamic certification requires rigorous adherence to standards and ongoing documentation, posing administrative challenges.

1.2 Biodynamic Farm at Rodale Institute, USA

Overview: Rodale Institute's biodynamic farm in Pennsylvania is a leading research and education centre for organic and biodynamic farming. Established in 1947, the institute focuses on advancing sustainable agriculture through research, education, and outreach.

Key Features

- **Research Focus:** The biodynamic farm at Rodale Institute conducts research on soil health, crop management, and the impact of biodynamic practices on farm productivity and sustainability.
- **Educational Programs:** The institute offers training programs and workshops for farmers, researchers, and students, promoting the adoption of biodynamic and organic practices.

- **Community Outreach:** Rodale Institute actively engages with the broader agricultural community through conferences, publications, and collaborations.

Successes

- **Innovative Research:** The institute has made significant contributions to understanding the benefits and challenges of biodynamic farming through its research programs.
- **Farmer Support:** Rodale Institute has supported numerous farmers in adopting biodynamic and organic practices through its educational and outreach efforts.

Challenges

- **Funding and Resources:** Securing funding for research and education programs is a continual challenge, requiring ongoing support from donors and partners.

1.3 La Ferme du Bec Hellouin, France

Overview: La Ferme du Bec Hellouin, located in Normandy, France, is known for its intensive biodynamic farming practices and its focus on permaculture principles. The farm is a demonstration site for sustainable agriculture and a model for integrating biodiversity and productivity.

Key Features

- **Permaculture Integration:** The farm combines biodynamic farming with permaculture principles, creating a diverse and resilient farming system.
- **Research and Demonstration:** La Ferme du Bec Hellouin serves as a research and demonstration site, showcasing innovative farming techniques and promoting biodiversity.
- **Educational Workshops:** The farm offers workshops and training sessions for farmers, students, and enthusiasts, emphasizing hands-on learning and practical applications.

Successes

- **High Productivity:** The farm has achieved impressive levels of productivity and sustainability through its integrated approach to biodynamic and permaculture practices.
- **Global Recognition:** La Ferme du Bec Hellouin has gained international recognition for its contributions to sustainable agriculture and its role as a demonstration site.

Challenges

- **Economic Viability:** Balancing economic viability with ecological principles remains a challenge, requiring careful management of resources and income streams.

2. Startup Needs for a Biodynamic Farm

2.1 Land Selection and Preparation

Choosing the Right Location

- **Soil Quality:** Select land with healthy, well-draining soil. Soil tests can help assess fertility and suitability for biodynamic practices.
- **Climate Conditions:** Consider local climate conditions, including temperature, rainfall, and sunlight, as these factors affect crop growth and farming practices.

Land Preparation

- **Soil Enrichment:** Implement practices to improve soil health, such as composting, cover cropping, and the use of biodynamic preparations.
- **Infrastructure:** Develop necessary infrastructure, including irrigation systems, composting facilities, and storage areas.

2.2 Certification and Standards

Understanding Certification Requirements:

- **Biodynamic Certification:** Obtain certification from recognized bodies such as Demeter or Biodynamic Association. This involves adhering to specific standards and practices outlined by the certification organization.
- **Documentation and Record-Keeping:** Maintain detailed records of farming practices, inputs, and compliance with certification standards.

Meeting Standards

- **Adherence to Principles:** Follow biodynamic principles, including the use of biodynamic preparations, lunar planting cycles, and sustainable resource management.
- **Regular Inspections:** Prepare for regular inspections by certification organizations to ensure ongoing compliance with biodynamic standards.

2.3 Initial Investments and Resources

Financial Planning

- **Budgeting:** Develop a detailed budget covering startup cost, including land acquisition, infrastructure development, equipment, and initial inputs.

- **Funding Sources:** Explore funding options such as grants, loans, and investor support. Consider applying for agricultural grants or subsidies that support sustainable farming practices.

Resource Acquisition

- **Biodynamic Inputs:** Source high-quality biodynamic inputs, including compost, preparations, and seeds. Establish relationships with suppliers who provide biodynamic and organic products.
- **Equipment and Tools:** Invest in essential equipment and tools for farming operations, including planting machinery, irrigation systems, and soil testing equipment.

2.4 Knowledge and Skills

Education and Training

- **Technical Knowledge:** Acquire knowledge and skills related to biodynamic farming practices through courses, workshops, and mentorship programs.
- **Hands-On Experience:** Gain practical experience by working on biodynamic farms or participating in field-based training programs.

Networking and Support

- **Join Organizations:** Become a member of biodynamic farming organizations and networks to access resources, support, and information.
- **Engage with Experts:** Seek advice and guidance from experienced biodynamic farmers and practitioners.

Conclusion

Biodynamic farming offers a holistic and sustainable approach to agriculture, with notable farms around the world exemplifying its principles and practices. Profiles of farms such as Demeter Farm, Rodale Institute, and La Ferme du Bec Hellouin highlight the successes and innovations possible within biodynamic agriculture. These farms demonstrate the potential for enhancing soil health, community engagement, and global recognition through biodynamic practices.

For those interested in starting a biodynamic farm, understanding startup needs is crucial. Key considerations include land selection, certification requirements, initial investments, and acquiring the necessary knowledge and skills. By addressing these needs and leveraging the experiences of established biodynamic farms, aspiring farmers can embark on a successful and sustainable journey in biodynamic agriculture.

Through careful planning, adherence to principles, and engagement with the broader biodynamic community, new farms can contribute to the growth and development of biodynamic farming, supporting a more sustainable and resilient agricultural future.

References

Groh, T., & McFadden, S. (1998). Farms of tomorrow revisited. SteinerBooks.

Koepf, H. H. (2005). The Biodynamic Farm: Agriculture in Service of the Earth and Humanity. SteinerBooks.

Patel, R. S. (2021). Biodynamic agriculture: a form of alternative agriculture. Agriculture\& food: e-newsletter.

Ravisankar, N., Panwar, A. S., & Dutta, D. ORGANIC FARMING-STATUS AND ASPECTS. Modern Concepts and Practices of Organic Farming for Safe Secured and Sustainable Food Production, 16.

13

Biodynamic Farming Around the World and International Organizations and Movements

Abstract

Biodynamic farming, rooted in the early 20th-century spiritual insights of Rudolf Steiner, has evolved into a globally recognized approach to agriculture that emphasizes the interconnectedness of soil, plants, animals, and cosmic rhythms. This article explores the principles and practices of biodynamic farming and traces its development across different regions of the world. It examines the contributions of international organizations and movements in promoting biodynamic practices, enhancing sustainability, and advocating for ecological harmony. The article also delves into the challenges and opportunities faced by biodynamic farmers globally and assesses the impact of biodynamic farming on soil health, crop quality, and community well-being. By highlighting case studies and initiatives from various countries, the article aims to provide a comprehensive understanding of the global biodynamic farming movement and its role in shaping the future of sustainable agriculture.

Introduction

Biodynamic farming is more than just a method of organic farming; it represents a holistic approach to agriculture that considers the farm as a self-sustaining ecosystem. Originating from the teachings of Rudolf Steiner in the 1920s, biodynamic farming integrates spiritual, ethical, and ecological principles, emphasizing the interrelationship between soil, plants, animals, and cosmic forces. This article aims to explore the global spread of biodynamic farming, the role of international organizations in its promotion, and the movements that have arisen to support its principles and practices.

1. Historical Background and Foundations of Biodynamic Farming

The roots of biodynamic farming can be traced back to Rudolf Steiner's 1924 lectures, which laid the foundation for this innovative approach to agriculture. Steiner's insights were driven by concerns over the degradation of soil fertility and the impact of industrial agriculture on the environment. His teachings led

DOI: 10.1201/9781042015597-13

to the development of biodynamic practices that focus on the farm as a closed system, relying on composting, crop rotation, and specific preparations to enhance soil and plant health.

2. Core Principles of Biodynamic Farming

Biodynamic farming is a holistic and regenerative approach to agriculture that views the farm as a self-sustaining organism. It builds upon organic farming practices by incorporating spiritual and cosmic dimensions into farming. Developed by Austrian philosopher Rudolf Steiner in the 1920s, biodynamic farming integrates soil fertility, plant growth, and animal care into a unified, self-nourishing system. This approach emphasizes the interconnectedness of life and the farm's relationship with the wider cosmos.

The core principles of biodynamic farming are rooted in these foundational ideas:

2.1 The Farm as a Self-Sustaining Organism

At the heart of biodynamic farming is the concept of the farm as a self-sustaining, living organism. This principle emphasizes the integration of all components of the farm—soil, plants, animals, and humans—into a balanced and harmonious whole.

- **Diversity and Interdependence:** Biodynamic farms prioritize diversity in crops and livestock to create a self-sustaining system. Each component of the farm plays a specific role, supporting the health and vitality of the entire organism. For example, livestock provide manure for compost, which in turn enhances soil fertility and supports plant growth. The plants, in turn, provide feed for the animals, completing the cycle.
- **Internal Nutrient Cycles:** A biodynamic farm aims to minimize external inputs by maintaining internal nutrient cycles. This involves composting organic waste, using cover crops to fix nitrogen, and integrating livestock to recycle nutrients through manure. The goal is to create a closed-loop system where the farm generates its fertility and resilience.
- **Local Adaptation:** Biodynamic farming practices are tailored to the specific conditions of the farm, including its soil, climate, and ecological context. Farmers observe and respond to the unique needs of their land, fostering a deep connection with the local environment.

2.2 Biodynamic Preparations

Biodynamic preparations are unique to this farming method and are essential for enhancing soil fertility, plant growth, and overall farm vitality. These preparations are made from natural materials like herbs, minerals, and animal products and are applied in homeopathic doses to the farm.

- **BD 500 (Horn Manure):** Made by filling a cow horn with cow manure and burying it in the soil over the winter. The manure undergoes a transformation, and the resulting preparation is used to enhance soil structure and microbial activity. It is typically applied as a spray to the soil to stimulate root growth and nutrient absorption.
- **BD 501 (Horn Silica):** Prepared by filling a cow horn with finely ground quartz and burying it in the soil during the summer. The preparation is then sprayed on crops to improve photosynthesis, increase resistance to diseases, and enhance the quality of fruits and vegetables.
- **Compost Preparations (BD 502-507):** These include preparations made from yarrow, chamomile, stinging nettle, oak bark, dandelion, and valerian. These are added to compost piles to aid decomposition and enhance the compost's ability to support plant health.
- **BD 508 (Equisetum or Horsetail):** Used to control fungal diseases in crops. It is prepared from the horsetail plant and applied as a spray to strengthen plants' defenses against fungal infections.

2.3 Cosmic and Rhythmic Influences

Biodynamic farming recognizes the influence of cosmic and lunar rhythms on plant and animal life. This principle is based on the idea that celestial bodies, such as the moon, planets, and stars, exert subtle but significant effects on life on Earth.

- **Lunar Rhythms:** The phases of the moon are believed to affect plant growth, and biodynamic farmers align their planting, cultivating, and harvesting activities with the lunar cycle. For example, planting seeds during the waxing moon is thought to enhance germination and growth, while the waning moon is considered a good time for pruning and harvesting.
- **Zodiacal Influences:** The position of the moon and planets relative to the zodiac constellations is also considered when planning farm activities. The biodynamic calendar divides the month into periods favorable for different types of plant growth (root, leaf, flower, and fruit), depending on the moon's position in the zodiac.
- **Seasonal Rhythms:** Biodynamic farming aligns with the natural rhythms of the seasons, emphasizing the importance of timing in agricultural activities. Preparations are made and applied at specific times of the year

to align with these rhythms, ensuring that the farm is in harmony with the broader natural cycles.

2.4 Holistic Health and Vitality

Biodynamic farming places a strong emphasis on the health and vitality of the entire farm system, including soil, plants, animals, and humans. This holistic approach seeks to cultivate vitality at every level of the farm ecosystem.

- **Soil Health:** The vitality of the soil is central to biodynamic farming. Healthy soil is rich in organic matter, teeming with microbial life, and capable of sustaining robust plant growth. Biodynamic practices, such as composting, crop rotation, and the use of biodynamic preparations, are designed to enhance soil fertility and structure.
- **Plant Vitality:** Biodynamic farming seeks to produce crops that are not only free from synthetic chemicals but also rich in vitality. This means that plants are grown in healthy, living soil and are nurtured in a way that enhances their nutritional value, flavor, and resistance to pests and diseases.
- **Animal Welfare:** Animals on biodynamic farms are treated with care and respect, reflecting their integral role in the farm ecosystem. They are provided with natural diets, ample space, and living conditions that allow them to express their natural behaviors. In return, animals contribute to the farm's fertility and balance.
- **Human Well-Being:** Biodynamic farming recognizes the farmer's role as a steward of the land, responsible for nurturing the farm organism. The well-being of the farmer and farm workers is considered essential, and the farming practices are designed to foster a deep connection between humans and the land they cultivate.

2.5 Spiritual Connection to the Land

Biodynamic farming acknowledges the spiritual dimension of agriculture, encouraging farmers to cultivate a deep, mindful relationship with their land.

- **Mindful Observation:** Biodynamic farmers are encouraged to observe their farm's natural processes and rhythms closely. This mindfulness allows them to respond intuitively to the needs of their crops, animals, and soil, fostering a deep connection to the land.
- **Ritual and Intention:** Some biodynamic practices include rituals, such as the stirring of preparations or the timing of planting activities, that are performed with a conscious intention to support the farm's health and

vitality. These rituals reflect a belief in the interconnectedness of all life and the importance of intentionality in farming practices.

- **Community and Stewardship:** Biodynamic farming often extends beyond the individual farm to include a sense of community and stewardship for the broader environment. Farmers may engage with local communities, share knowledge, and participate in cooperative efforts to promote sustainable and regenerative agriculture.

3. Global Spread of Biodynamic Farming

Since its inception in Europe, biodynamic farming has spread to various parts of the world, adapting to different climates, cultures, and agricultural challenges. The global spread of biodynamic farming can be understood through the examination of key regions where it has taken root:

Europe

Europe remains the heartland of biodynamic farming, with countries like Germany, Austria, and Switzerland leading in both the number of biodynamic farms and the development of biodynamic certification standards. The Demeter International certification, established in 1928, plays a crucial role in maintaining the integrity of biodynamic practices across Europe.

North America

In North America, biodynamic farming has gained traction particularly in the United States and Canada. The movement has been supported by organizations such as the Biodynamic Association, which provides education, research, and certification services. The adoption of biodynamic practices in vineyards, particularly in California, has brought significant attention to the method's benefits in producing high-quality, terroir-driven wines.

Latin America

Latin America has seen a growing interest in biodynamic farming, particularly in countries like Brazil, Argentina, and Chile. In these regions, biodynamic practices are being integrated into coffee, wine, and tropical fruit production. The emphasis on sustainable farming practices aligns with the region's rich biodiversity and the need for resilient agricultural systems.

Asia

In Asia, biodynamic farming is gradually gaining recognition, especially in India, China, and Japan. Indian biodynamic farmers, for instance, are focusing on reviving traditional agricultural practices that align with the principles of biodynamics. Organizations like the Biodynamic Association of India are playing a key role in promoting and educating farmers about biodynamic methods.

Africa

In Africa, biodynamic farming is being adopted as part of broader efforts to promote sustainable agriculture and food security. Countries like South Africa and Egypt have established biodynamic farms that serve as models for integrating these practices into diverse agricultural landscapes. These initiatives are often supported by international organizations and collaborations with European biodynamic practitioners.

Australia and New Zealand

Australia and New Zealand have also embraced biodynamic farming, particularly in the viticulture sector. The region's emphasis on high-quality, sustainable food production has led to a growing number of biodynamic farms, supported by organizations like Biodynamic Agriculture Australia.

4. Role of International Organizations in Promoting Biodynamic Farming

Several international organizations have been instrumental in the promotion and development of biodynamic farming worldwide. These organizations provide education, certification, and support to farmers, helping to spread the principles and practices of biodynamics globally.

i. **Demeter International:** Demeter International is the oldest and most recognized certification body for biodynamic farming. Established in 1928, it sets rigorous standards for biodynamic farming practices and certifies farms and products that meet these criteria. Demeter's work is crucial in maintaining the integrity of biodynamic farming and ensuring that consumers can trust the products they purchase.

ii. **Biodynamic Federation Demeter International:** The Biodynamic Federation Demeter International (BFDI) was formed in 2020 as a global alliance of biodynamic organizations. It aims to unify and strengthen the biodynamic movement by coordinating efforts across different regions, providing resources and support to local biodynamic initiatives, and representing the movement in international forums.

iii. **The International Biodynamic Association (IBDA):** The International Biodynamic Association (IBDA) serves as a global network for biodynamic organizations and practitioners. It fosters collaboration, research, and knowledge exchange among its members, helping to advance biodynamic farming practices and principles worldwide.

iv. **The Biodynamic Association:** The Biodynamic Association, based in the United States, is a key organization in the North American biodynamic movement. It provides educational resources, organizes conferences and workshops, and supports research on biodynamic farming. The association also plays a significant role in advocating for policies that support biodynamic and organic agriculture.

5. Biodynamic Movements and Initiatives Around the World

Biodynamic farming has inspired numerous movements and initiatives that aim to promote sustainable agriculture, ecological balance, and social justice. These movements often work in collaboration with international organizations to advance the goals of the biodynamic movement globally.

i. **The Slow Food Movement:** The Slow Food Movement, founded in Italy in 1986, shares many of the same values as the biodynamic movement, particularly the emphasis on sustainable, locally sourced food. Biodynamic farmers often collaborate with Slow Food initiatives to promote agricultural practices that protect the environment, preserve biodiversity, and support local communities.

ii. **The Organic Agriculture Movement:** The organic agriculture movement and the biodynamic movement have a shared history and continue to work closely together. Many biodynamic farms are also certified organic, and the principles of organic farming are integral to biodynamic practices. Organizations like the International Federation of Organic Agriculture Movements (IFOAM) often include biodynamic farming as part of their broader mission to promote sustainable agriculture.

iii. **Community-Supported Agriculture (CSA):** Community-Supported Agriculture (CSA) is a model that has been embraced by biodynamic farmers around the world. In a CSA, consumers purchase a share of a farm's harvest in advance, providing financial support to the farmer and receiving a regular supply of fresh, seasonal produce. This model fosters a direct connection between farmers and consumers and supports the principles of sustainability and community engagement that are central to biodynamic farming.

iv. **The Global Ecovillage Network (GEN):** The Global Ecovillage Network (GEN) is an international organization that promotes the development of sustainable communities, or ecovillages, around the world. Biodynamic farming practices are often integral to the agricultural systems within these ecovillages, as they align with the values of ecological sustainability, community resilience, and holistic living.

6. Challenges and Opportunities in Biodynamic Farming

While biodynamic farming offers numerous benefits, it also faces challenges that need to be addressed to ensure its continued growth and success.

a. **Challenges**

1. **Certification and Standards**: The rigorous standards required for biodynamic certification can be a barrier for small-scale farmers, particularly in developing countries. The cost and complexity of certification may limit access to the biodynamic market for these farmers.
2. **Education and Awareness**: There is a need for greater education and awareness about biodynamic farming, both among farmers and consumers. Misconceptions about the spiritual aspects of biodynamics can also be a barrier to its wider adoption.
3. **Market Access**: Biodynamic products often command a premium price, which can limit their accessibility to a broader market. Farmers may struggle to find markets for their products, particularly in regions where awareness of biodynamic farming is low.
4. **Research and Development**: While there is growing interest in biodynamic farming, more research is needed to quantify its benefits and address the challenges it faces. This includes research into the effectiveness of biodynamic preparations, the impact of cosmic rhythms on plant growth, and the long-term sustainability of biodynamic practices.

b. **Opportunities**

1. **Global Expansion**: There is significant potential for the global expansion of biodynamic farming, particularly in regions where sustainable agriculture is a priority. International organizations and movements can play a key role in supporting this expansion through education, certification, and market development.
2. **Integration with Other Movements**: Biodynamic farming can benefit from greater integration with other sustainable agriculture movements, such as permaculture, regenerative agriculture, and agroecology. These movements share similar goals and principles, and collaboration could lead to stronger, more resilient agricultural systems worldwide.
3. **Policy Advocacy**: Biodynamic organizations and movements have the opportunity to influence agricultural policy at both national and international levels. By advocating for policies that support sustainable

farming practices, such as subsidies for biodynamic farmers or incentives for organic certification, the biodynamic movement can help shape the future of agriculture.

4. **Consumer Demand**: As consumer awareness of environmental and health issues grows, there is increasing demand for sustainable and ethically produced food. Biodynamic farming, with its emphasis on ecological balance and holistic management, is well-positioned to meet this demand. Educating consumers about the benefits of biodynamic products can help expand the market and support the growth of biodynamic farming.
5. **Technological Advancements**: While biodynamic farming emphasizes traditional and natural methods, there is potential to integrate modern technology to enhance productivity and sustainability. For example, precision agriculture techniques could be used to optimize the application of biodynamic preparations, and digital platforms could help connect biodynamic farmers with consumers and markets.

7. Impact of Biodynamic Farming on Soil Health and Crop Quality

One of the core tenets of biodynamic farming is its focus on improving soil health and maintaining ecological balance. The long-term effects of biodynamic practices on soil fertility, microbial diversity, and overall ecosystem health are significant.

Biodynamic farming practices, such as the use of compost and biodynamic preparations, have been shown to enhance soil structure, increase organic matter content, and improve microbial activity. These improvements in soil health contribute to greater resilience against erosion, drought, and other environmental stresses. Studies have demonstrated that biodynamic soils tend to have higher levels of beneficial microorganisms and greater biodiversity compared to conventionally farmed soils.

Biodynamic farming not only aims to produce high yields but also to enhance the quality of the crops produced. Research has indicated that biodynamic crops often have higher levels of vitamins, minerals, and antioxidants compared to conventionally grown crops. The focus on soil health and the use of natural preparations contribute to the development of plants that are more resistant to pests and diseases, leading to healthier and more nutrient-dense produce.

8. Case Studies: Biodynamic Farming Around the World

To illustrate the global impact of biodynamic farming, this section presents case studies from various regions, highlighting the successes and challenges faced by biodynamic farmers.

Europe: The Sekem Initiative in Egypt

Founded in 1977 by Dr. Ibrahim Abouleish, the Sekem Initiative in Egypt is a pioneering example of biodynamic farming in a challenging environment. Sekem began as a small farm in the desert and has since grown into a thriving community that produces biodynamic crops, herbs, and medicinal plants. The initiative has also expanded to include schools, healthcare facilities, and research centers, all focused on sustainable development and social responsibility. Sekem's success demonstrates the potential of biodynamic farming to transform arid and degraded landscapes into productive, thriving ecosystems.

North America: Frey Vineyards in California

Frey Vineyards, located in California's Mendocino County, is a leading example of biodynamic viticulture. The Frey family has been practicing biodynamic farming since the 1980s, producing wines that are both organic and biodynamic. The vineyard's holistic approach to farming, which includes the integration of animals, composting, and the use of biodynamic preparations, has led to the production of high-quality wines that reflect the unique terroir of the region. Frey Vineyards is also a strong advocate for sustainable agriculture and has played a key role in promoting biodynamic practices in the wine industry.

Latin America: Biodynamic Coffee Farming in Brazil

Brazil, one of the world's largest coffee producers, has seen a growing interest in biodynamic coffee farming. One notable example is Fazenda Camocim, a biodynamic coffee farm in the Espirito Santo region. The farm practices biodynamic agriculture to produce high-quality coffee beans while preserving the natural environment. Biodynamic practices, such as composting, crop rotation, and the use of biodynamic preparations, have helped improve soil fertility, enhance coffee quality, and increase biodiversity on the farm. Fazenda Camocim's success has inspired other coffee producers in the region to adopt biodynamic methods.

Asia: Biodynamic Farming in India

In India, biodynamic farming has gained recognition as a method that aligns with traditional agricultural practices. Navdanya, an organization founded by environmental activist Dr. Vandana Shiva, has been instrumental in promoting biodynamic farming in India. Navdanya works with small-scale farmers to implement biodynamic practices, focusing on seed sovereignty, biodiversity conservation, and sustainable agriculture. The organization has established biodynamic farms across India, producing a wide range of crops, including grains, pulses, fruits, and vegetables. Navdanya's efforts have helped revive traditional farming practices and improve the livelihoods of small-scale farmers.

Africa: Spier Wine Farm in South Africa

Spier Wine Farm, located in South Africa's Stellenbosch region, is a leading example of biodynamic farming in Africa. The farm has been practicing biodynamic viticulture since the early 2000s, producing wines that are recognized for their quality and sustainability. Spier's commitment to biodynamic farming extends beyond the vineyard, as the farm also focuses on conservation efforts, water management, and community engagement. The integration of biodynamic practices with a strong focus on sustainability has made Spier a model for other farms in the region.

Australia: Powlett Hill Farm

Powlett Hill Farm, located in Victoria, Australia, is a biodynamic farm that produces grains, pulses, and sheep. The farm has been practicing biodynamic agriculture for over 20 years, focusing on soil health, biodiversity, and sustainable farming practices. Powlett Hill uses biodynamic preparations, crop rotation, and composting to maintain soil fertility and produce high-quality crops. The farm's commitment to biodynamic principles has resulted in healthy soils, diverse ecosystems, and a thriving farm business.

9. Future Prospects of Biodynamic Farming

As the world faces increasing environmental challenges, such as climate change, soil degradation, and loss of biodiversity, the principles and practices of biodynamic farming offer a promising path forward. The future of biodynamic farming will likely be shaped by several key factors:

1. **Climate Change and Biodynamic Farming:** Biodynamic farming's emphasis on soil health, biodiversity, and ecological balance makes it well-suited to address the challenges of climate change. Healthy soils with high organic matter content can sequester carbon, helping to mitigate the impact of greenhouse gas emissions. Biodynamic practices, such as crop rotation and the use of cover crops, can also enhance the resilience of farms to extreme weather events, such as droughts and floods.
2. **Integration with Agroecology and Regenerative Agriculture:** Biodynamic farming is increasingly being recognized as part of the broader movement towards agroecology and regenerative agriculture. These approaches share a common goal of creating sustainable and

resilient agricultural systems that work in harmony with nature. The integration of biodynamic practices with agroecological principles could enhance the effectiveness of both approaches and contribute to the development of more sustainable food systems.

3. **Education and Training:** The future growth of biodynamic farming will depend on the availability of education and training opportunities for farmers. Expanding access to biodynamic education, particularly in developing regions, will be crucial for the adoption of biodynamic practices on a larger scale. International organizations, such as the Biodynamic Federation Demeter International and the Biodynamic Association, will play a key role in providing resources, training, and support to farmers around the world.
4. **Research and Innovation:** Ongoing research and innovation will be essential for advancing biodynamic farming and addressing the challenges it faces. Areas of research could include the development of new biodynamic preparations, the exploration of the relationship between cosmic rhythms and plant growth, and the study of the long-term effects of biodynamic practices on soil health and crop quality. Collaborative research efforts between biodynamic organizations, universities, and research institutions will be important for generating new knowledge and improving biodynamic farming practices.
5. **Consumer Awareness and Market Development:** As consumer demand for sustainable and ethically produced food continues to grow, there will be opportunities for expanding the market for biodynamic products. Educating consumers about the benefits of biodynamic farming, such as its positive impact on soil health, biodiversity, and food quality, can help drive demand for biodynamic products. Developing strong markets for biodynamic products will also provide economic incentives for farmers to adopt biodynamic practices.

Conclusion

Biodynamic farming is a holistic and sustainable approach to agriculture that has the potential to address many of the environmental and social challenges facing the world today. By viewing the farm as a self-sustaining ecosystem, biodynamic farmers work in harmony with nature to produce healthy, high-quality food while preserving the environment for future generations. The global spread of biodynamic farming, supported by international organizations and movements, demonstrates its adaptability and relevance across different regions and cultures.

The future of biodynamic farming will depend on continued research, education, and collaboration among farmers, organizations, and consumers. As awareness of the importance of sustainable agriculture grows, biodynamic farming is poised to play an increasingly important role in shaping the future of food production. By embracing the principles of biodynamics, farmers around the world can contribute to a more sustainable and resilient agricultural system that benefits both people and the planet.

References

Geier, B. (2007). IFOAM and the history of the International Organic Movement. In Organic Farming: an international history (pp. 175-186). Wallingford UK: CABI.

Koepf, H. H. (2005). The Biodynamic Farm: Agriculture in Service of the Earth and Humanity. SteinerBooks.

Paull, J. (2010). From France to the world: The international federation of organic agriculture movements (IFOAM). Journal of Social Research & Policy, 1(2).

Staudenmaier, P. (2013). Organic farming in Nazi Germany: The politics of biodynamic agriculture, 1933–1945. Environmental History.

Wierzbicki, M. (2012). Biodynamic farming and nutrition and the conscious evolution of consumer society.

14

Future of Biodynamic Farming and Emerging Trends and Technologies

Abstract

The future of biodynamic farming is intricately linked to emerging trends and technologies that are transforming the landscape of agriculture. As the world grapples with environmental challenges, the demand for sustainable farming practices is intensifying. Biodynamic farming, rooted in ecological principles and holistic management, is gaining prominence as a viable solution for sustainable food production. This article explores the future of biodynamic farming, focusing on the emerging trends and technologies that are poised to shape its development. From advances in digital agriculture and precision farming to the integration of renewable energy and sustainable practices, the article examines how these innovations can enhance the effectiveness and scalability of biodynamic farming. Additionally, it highlights the role of research, education, and policy support in driving the adoption of biodynamic practices worldwide. By embracing these trends and technologies, biodynamic farming can contribute to a more resilient and sustainable global food system.

Introduction

Biodynamic farming, a holistic approach to agriculture that views the farm as a self-sustaining ecosystem, has been practiced for nearly a century. Based on the teachings of Rudolf Steiner, biodynamic farming emphasizes the use of natural preparations, crop rotation, composting, and the alignment of farming activities with cosmic rhythms. As the world faces unprecedented environmental challenges, such as climate change, soil degradation, and loss of biodiversity, the principles of biodynamic farming are becoming increasingly relevant.

In recent years, there has been growing interest in the potential of biodynamic farming to contribute to a more sustainable and resilient food system. This interest is being fueled by emerging trends and technologies that have the potential to enhance the effectiveness, scalability, and sustainability of biodynamic practices. As biodynamic farming continues to evolve, it is

DOI: 10.1201/9781042015597-14

essential to explore these emerging trends and technologies and understand how they can shape the future of this unique approach to agriculture.

Emerging Trends in Biodynamic Farming

Several key trends are emerging in the field of biodynamic farming, driven by the need for sustainable food production and the increasing demand for organic and ethically produced food. These trends are influencing the way biodynamic farming is practiced and are paving the way for its future development.

1. Integration of Digital Agriculture

Digital agriculture, which encompasses the use of technology and data to optimize farming practices, is becoming increasingly important in biodynamic farming. While biodynamic farming has traditionally emphasized natural and manual methods, the integration of digital tools can enhance the efficiency and precision of farming activities.

1.1 Precision Farming

Precision farming involves the use of data and technology to monitor and manage crops with greater accuracy. In biodynamic farming, precision farming can be used to optimize the application of biodynamic preparations, monitor soil health, and track crop growth. For example, sensors can be used to measure soil moisture, temperature, and nutrient levels, allowing farmers to make informed decisions about irrigation, fertilization, and planting. Drones and satellite imagery can also be used to monitor crop health and identify areas that require attention.

By integrating precision farming techniques, biodynamic farmers can improve the efficiency of their operations, reduce waste, and enhance crop quality. This approach aligns with the principles of biodynamic farming, which emphasize the importance of working in harmony with nature and minimizing the use of external inputs.

1.2 Farm Management Software

Farm management software is another emerging trend in biodynamic farming. These digital platforms allow farmers to track and manage various aspects of their operations, such as crop planning, inventory management, and financial records. For biodynamic farmers, farm management software can be used to document the use of biodynamic preparations, track crop rotation schedules, and monitor the overall health of the farm ecosystem.

The use of farm management software can also facilitate certification processes for biodynamic products, as it provides a transparent and organized record of farming practices. As consumer demand for transparency and traceability in food production grows, the adoption of farm management software can help biodynamic farmers meet these expectations and strengthen their market position.

2. Adoption of Renewable Energy

Renewable energy is playing an increasingly important role in sustainable agriculture, including biodynamic farming. As concerns about climate change and fossil fuel dependence grow, biodynamic farmers are turning to renewable energy sources to power their operations and reduce their environmental impact.

2.1 Solar Energy

Solar energy is one of the most widely adopted renewable energy sources in agriculture. Biodynamic farms can use solar panels to generate electricity for irrigation systems, farm equipment, and processing facilities. Solar energy is particularly well-suited to biodynamic farming, as it aligns with the principles of self-sufficiency and ecological sustainability. By harnessing solar energy, biodynamic farmers can reduce their reliance on non-renewable energy sources and lower their carbon footprint.

2.2. Wind Energy

Wind energy is another renewable energy source that can be utilized in biodynamic farming. Wind turbines can be installed on farms to generate electricity, which can be used to power various farming operations. In regions with consistent wind patterns, wind energy can provide a reliable and sustainable source of power for biodynamic farms.

2.3 Bioenergy

Bioenergy, derived from organic materials such as crop residues and animal manure, is also gaining traction in biodynamic farming. Biodynamic farms can produce bioenergy through the process of anaerobic digestion, which breaks down organic matter to produce biogas. This biogas can be used as a renewable energy source for heating, electricity generation, and cooking. The use of bioenergy not only provides a sustainable energy solution but also helps manage farm waste and reduce greenhouse gas emissions.

3. Focus on Soil Health and Regenerative Practices

Soil health is a central focus of biodynamic farming, and emerging trends in regenerative agriculture are closely aligned with biodynamic principles. Regenerative agriculture aims to restore and enhance soil health through practices that promote biodiversity, improve water retention, and increase organic matter content. Biodynamic farming, with its emphasis on composting, crop rotation, and the use of natural preparations, is inherently regenerative.

3.1 Composting and Soil Amendments

Composting is a key practice in biodynamic farming, as it helps build healthy, nutrient-rich soil. Emerging trends in composting include the use of advanced composting techniques, such as vermicomposting and thermophilic composting, which accelerate the decomposition process and produce high-quality compost. Biodynamic farmers are also exploring the use of biochar, a form of charcoal that can be added to soil to improve its structure and increase its ability to retain nutrients and water.

3.2 Cover Cropping and Crop Rotation

Cover cropping and crop rotation are regenerative practices that are gaining prominence in biodynamic farming. Cover crops, such as legumes and grasses, are planted between main crops to protect the soil, prevent erosion, and fix nitrogen. Crop rotation involves alternating different types of crops on the same land to disrupt pest and disease cycles, improve soil fertility, and enhance biodiversity. These practices align with biodynamic principles and contribute to the long-term health and productivity of the farm ecosystem.

3.3. No-Till Farming

No-till farming, a practice that involves minimal soil disturbance, is another trend that is gaining traction in biodynamic farming. By reducing or eliminating tillage, biodynamic farmers can preserve soil structure, protect soil microbes, and reduce erosion. No-till farming also helps sequester carbon in the soil, contributing to climate change mitigation. The adoption of no-till practices in biodynamic farming represents a shift towards more sustainable and resilient agricultural systems.

4. Expansion of Organic and Biodynamic Certification

As consumer demand for organic and biodynamic products continues to grow, there is increasing interest in certification programs that verify the authenticity and sustainability of these products. Biodynamic certification, often overseen by organizations such as Demeter International, ensures that farms adhere to the strict principles and practices of biodynamic farming.

4.1 Global Expansion of Biodynamic Certification

The expansion of biodynamic certification to new regions and markets is an emerging trend in the future of biodynamic farming. As more consumers become aware of the benefits of biodynamic products, there is a growing demand for certified products in both developed and developing countries. This expansion presents opportunities for biodynamic farmers to access new markets and increase the visibility of biodynamic products on a global scale.

4.2 Digital Certification and Traceability

The use of digital technologies in certification and traceability is another emerging trend in biodynamic farming. Blockchain technology, for example, can be used to create transparent and tamper-proof records of biodynamic farming practices, from production to distribution. This level of traceability not only enhances consumer trust but also helps protect the integrity of biodynamic certification. As digital certification and traceability systems become more widespread, biodynamic farmers will be better equipped to meet the growing demand for transparency in food production.

5. Emerging Technologies in Biodynamic Farming

In addition to emerging trends, several new technologies are poised to transform the practice of biodynamic farming. These technologies, while still in the early stages of adoption, have the potential to enhance the effectiveness and sustainability of biodynamic farming practices.

5.1 Advanced Soil Testing and Monitoring

Soil health is a cornerstone of biodynamic farming, and advanced soil testing and monitoring technologies are providing new insights into soil quality and fertility. These technologies enable biodynamic farmers to make more informed decisions about soil management and crop production.

a. **Soil Sensors and Remote Sensing:** Soil sensors and remote sensing technologies allow farmers to monitor soil conditions in real-time. These sensors can measure key soil parameters, such as moisture levels, nutrient content, pH, and temperature. By continuously monitoring soil health, biodynamic farmers can optimize the use of biodynamic preparations, adjust irrigation practices, and improve crop management.

   Remote sensing technologies, such as satellite imagery and drones, can also be used to assess soil health across large areas of farmland. These technologies provide detailed maps of soil conditions, helping farmers identify areas that require attention and enabling more precise application of biodynamic practices.

b. **Microbial Analysis:** Advances in microbial analysis are providing biodynamic farmers with a deeper understanding of the soil microbiome—the diverse community of microorganisms that play a crucial role in soil health and fertility. By analyzing the composition and function of soil microbes, biodynamic farmers can tailor their practices to support beneficial microbial communities and enhance soil health. Techniques such as DNA sequencing and metagenomics allow for the identification of specific microbial species and their roles in nutrient cycling, disease suppression, and organic matter decomposition. This knowledge can inform the use of biodynamic preparations and other soil amendments, leading to more effective and targeted interventions.

5.2 Automation and Robotics in Biodynamic Farming

Automation and robotics are emerging technologies that have the potential to revolutionize biodynamic farming. While biodynamic farming traditionally emphasizes manual labor and hands-on management, the integration of automated systems can improve efficiency and reduce labor costs, allowing farmers to focus more on the holistic management of the farm ecosystem.

a. **Automated Weeding and Cultivation:** Robotic systems for weeding and cultivation are becoming increasingly sophisticated, offering biodynamic farmers a way to manage crops with minimal soil disturbance. These robots can identify and remove weeds without the use of chemical herbicides, aligning with the principles of biodynamic farming. Automated cultivation systems can also be used to prepare soil and plant crops with precision, ensuring optimal growing conditions and reducing the need for manual labor.

b. **Harvesting and Processing Automation:** Automation is also making its way into the harvesting and processing stages of biodynamic farming. Automated harvesting machines can reduce the time and labor required to bring crops to market, while also minimizing damage to the produce. In processing facilities, automation can streamline tasks such as sorting, packaging, and quality control, allowing biodynamic farms to operate more efficiently and maintain high standards of product quality.

5.3 Integration of Artificial Intelligence (AI) and Machine Learning

Artificial intelligence (AI) and machine learning are powerful tools that can enhance decision-making and problem-solving in biodynamic farming. By analyzing vast amounts of data from sensors, weather forecasts, and other sources, AI can provide biodynamic farmers with actionable insights that improve farm management and sustainability.

a. **Predictive Analytics:** Predictive analytics, powered by AI and machine learning, can help biodynamic farmers anticipate challenges and optimize their practices. For example, AI can analyze historical weather data and soil conditions to predict the best times for planting, applying biodynamic preparations, and harvesting. Predictive models can also forecast potential pest outbreaks or disease issues, allowing farmers to take preventive measures and reduce crop losses.

b. **AI-Driven Farm Management:** AI-driven farm management platforms can integrate data from various sources, including soil sensors, satellite imagery, and market trends, to provide biodynamic farmers with comprehensive insights into their operations. These platforms can recommend specific actions to improve crop yields, enhance soil health, and increase the overall sustainability of the farm. By leveraging AI, biodynamic farmers can make more informed decisions that align with the principles of ecological and sustainable farming.

5.4 Advancements in Biodynamic Preparations

Biodynamic preparations, made from natural materials such as herbs, minerals, and animal products, are a key component of biodynamic farming. Recent advancements in the formulation and application of these preparations are enhancing their effectiveness and expanding their use in biodynamic farming.

a. **Enhanced Formulation Techniques:** Researchers and practitioners are exploring new ways to enhance the formulation of biodynamic preparations. For example, the use of fermentation techniques can increase the potency and bioavailability of the active ingredients in these preparations. Innovations in mixing and aging processes are also being developed to improve the stability and efficacy of biodynamic preparations.

b. **Microbial Inoculants:** The use of microbial inoculants, which introduce beneficial microbes into the soil, is gaining attention in biodynamic farming. These inoculants can work in synergy with biodynamic preparations to enhance soil health, improve nutrient availability, and promote plant growth. By harnessing the power of beneficial microbes, biodynamic farmers can further strengthen the resilience and productivity of their farm ecosystems.

6. The Role of Education, Research, and Policy in the Future of Biodynamic Farming

The future of biodynamic farming will be shaped not only by emerging trends and technologies but also by the support provided through education, research, and policy initiatives. These elements are crucial for the continued growth and development of biodynamic farming on a global scale.

6.1 Education and Training for Biodynamic Farmers

As biodynamic farming evolves, there is a growing need for education and training programs that equip farmers with the knowledge and skills required to implement new practices and technologies. Educational institutions, non-governmental organizations (NGOs), and biodynamic associations play a vital role in providing this training.

a. **Farmer-to-Farmer Networks:** One effective approach to education in biodynamic farming is the establishment of farmer-to-farmer networks. These networks facilitate the exchange of knowledge and best practices among biodynamic farmers, allowing them to learn from each other's experiences and innovations. By participating in these networks, farmers can stay informed about emerging trends and technologies and adopt new practices that enhance the sustainability of their operations.

b. **Formal Education Programs:** Formal education programs, such as degree courses and vocational training, are also important for the future of biodynamic farming. These programs can cover topics such as soil science, organic farming techniques, sustainable energy, and digital agriculture. By providing a comprehensive education in biodynamic farming, these programs can prepare the next generation of farmers to lead the way in sustainable agriculture.

6.2 Research and Development in Biodynamic Farming

Research and development (R&D) are essential for advancing biodynamic farming practices and technologies. Continued investment in R&D can lead to new innovations that improve the efficiency, sustainability, and resilience of biodynamic farming.

a. **Collaborative Research Initiatives:** Collaborative research initiatives that bring together farmers, scientists, and policymakers are key to driving innovation in biodynamic farming. These initiatives can explore topics such as soil health, climate change adaptation, and the development of new biodynamic preparations. By working together, stakeholders can identify challenges and opportunities in biodynamic farming and develop solutions that benefit the entire farming community.

b. **Research on Biodynamic Preparations:** Research on the effectiveness and mechanisms of biodynamic preparations is another important area of focus. By gaining a deeper understanding of how these preparations work, scientists can refine their formulation and application, making them more effective and accessible to farmers. This research can also provide empirical evidence to support the use of biodynamic preparations, helping to build credibility and trust among consumers and policymakers.

6.3 Policy Support for Biodynamic Farming

Policy support is crucial for the widespread adoption and success of biodynamic farming. Governments and international organizations can play a significant role in promoting biodynamic farming through policies that incentivize sustainable practices and provide financial support to farmers.

a. **Subsidies and Incentives:** Subsidies and incentives for biodynamic farming can encourage more farmers to adopt these practices. For example, governments can offer financial incentives for the use of renewable energy, the adoption of precision farming technologies, or the certification of biodynamic products. These incentives can help offset the initial costs of implementing new practices and technologies, making biodynamic farming more accessible to small and medium-sized farms.

b. **Regulatory Frameworks:** The development of clear regulatory frameworks for biodynamic farming is also important. These frameworks can establish standards for biodynamic certification, ensure the integrity of biodynamic products, and protect the interests of farmers and consumers. By creating a supportive regulatory environment, governments can foster the growth of the biodynamic farming sector and contribute to the development of a sustainable global food system.

Conclusion

The future of biodynamic farming is bright, with emerging trends and technologies offering new opportunities to enhance the sustainability, efficiency, and resilience of this holistic approach to agriculture. By integrating digital tools, adopting renewable energy, focusing on soil health, and embracing automation and AI, biodynamic farmers can continue to lead the way in sustainable food production. Education, research, and policy support will play a critical role in driving the adoption of these innovations and ensuring the continued growth of biodynamic farming worldwide. As the world faces increasing environmental challenges, biodynamic farming will remain a vital and evolving component of the global effort to build a more sustainable and resilient food system.

References

Paull, J., & Hennig, B. (2020). A world map of biodynamic agriculture. Agricultural and Biological Sciences Journal, 6(2), 114-119.

Ravisankar, N., Panwar, A. S., & Dutta, D. ORGANIC FARMING-STATUS AND ASPECTS. Modern Concepts and Practices of Organic Farming for Safe Secured and Sustainable Food Production, 16.

Reijntjes, C., Haverkort, B., & Waters-Bayer, A. (1992). Farming for the Future. Macmillan Educ..

Selvi, S., Karthikeyan, R., & Vanitha, U. (2012, March). Organic farming: Technology for environment-friendly agriculture. In IEEE-International Conference On Advances In Engineering, Science And Management (ICAESM-2012) (pp. 132-136). IEEE.

Sharma, R., Aravind, T., & Sharma, R. (2019). Sustainable agriculture: Trends and opportunities for 21st Century. Journal of Applied and Natural Science, 11(3), 666-672.

Shitole, V., Desai, P., Inamdar, V., Waghulkar, S., & Mulla, T. (2024). The Revival Trends in the Wine Industry: A Comprehensive Literature Review. NATURALISTA CAMPANO, 28(1), 2958-2966.

15

Getting Started with Biodynamic Farming

Abstract

Biodynamic farming is a holistic and sustainable approach to agriculture that views the farm as a self-sustaining ecosystem. This article provides a comprehensive guide for beginners looking to embark on biodynamic farming, covering essential principles, practices, and preparations. It begins with an exploration of the philosophical underpinnings of biodynamic farming, rooted in the teachings of Rudolf Steiner, before delving into practical aspects such as soil health, crop rotation, composting, and pest management. The article also discusses the significance of biodynamic preparations, celestial planting calendars, and certification processes. With a focus on both traditional practices and modern adaptations, this guide aims to equip new biodynamic farmers with the knowledge and tools they need to start their journey toward a more sustainable and regenerative farming system.

Introduction

Biodynamic farming is more than just a method of organic farming; it is a holistic approach that views the farm as a living organism. This concept, developed by Austrian philosopher Rudolf Steiner in the early 20th century, integrates spiritual, ecological, and agricultural practices. Unlike conventional farming, which often focuses on maximizing yield through chemical inputs, biodynamic farming seeks to create a balanced and self-sustaining farm ecosystem. This approach emphasizes the importance of soil health, biodiversity, and the alignment of agricultural practices with natural rhythms.

Getting started with biodynamic farming requires an understanding of its core principles and a commitment to sustainable and regenerative practices. This article will guide you through the foundational elements of biodynamic farming, from understanding the philosophical basis to implementing practical techniques on your farm.

1. The Philosophical Foundations of Biodynamic Farming

Biodynamic farming is deeply rooted in the philosophical and spiritual teachings of Rudolf Steiner, an Austrian philosopher, social reformer, and

DOI: 10.1201/9781042015597-15

esotericist. His ideas form the backbone of the biodynamic approach, which goes beyond mere organic farming to incorporate a holistic understanding of nature and the farm as a living, interconnected system.

1.1 The Teachings of Rudolf Steiner

Rudolf Steiner's contribution to agriculture came through his series of eight lectures delivered in 1924, known as the "Agriculture Course." These lectures were a response to concerns raised by European farmers about the degradation of soil fertility and the decline in crop and livestock quality due to the increasing use of chemical fertilizers and pesticides. Steiner's lectures laid the groundwork for what would become biodynamic farming, emphasizing a return to natural and spiritual principles.

Steiner proposed that the health of the farm ecosystem is intricately linked to cosmic and earthly forces. He believed that these forces influence the vitality of plants, animals, and the soil. According to Steiner, a farm should be viewed as a self-sustaining organism, where every component—soil, plants, animals, and humans—plays a crucial role in maintaining balance and harmony.

Steiner introduced several key concepts that are central to biodynamic farming:

- **Anthroposophy:** Steiner's broader spiritual philosophy, known as Anthroposophy, underpins biodynamic farming. Anthroposophy is a spiritual science that seeks to understand the spiritual world and its relationship with the physical world. In the context of farming, it encourages farmers to develop a deep, intuitive connection with their land and to work in harmony with the natural rhythms of the earth and cosmos.
- **Biodynamic Preparations:** Steiner emphasized the use of specific preparations made from herbs, minerals, and animal products to enhance the vitality of the soil and crops. These preparations are applied in minute quantities but are believed to have powerful effects on the farm's energy and health. For example, BD 500 (horn manure) is prepared by burying cow horns filled with manure over the winter, and BD 501 (horn silica) involves burying ground quartz in cow horns during the summer. These preparations are then diluted and sprayed on fields to enhance soil fertility and plant vitality.
- **Cosmic Rhythms:** Steiner taught that the farm should operate in harmony with cosmic rhythms, particularly the lunar and planetary cycles. He believed that the positions of celestial bodies influence plant growth, animal behavior, and the overall health of the farm. This idea led to the development of the biodynamic planting calendar, which guides

farmers on the optimal times for planting, cultivating, and harvesting based on the moon and planets.

- **Spiritual Science:** Steiner's concept of spiritual science emphasizes the integration of scientific knowledge with spiritual insights. In biodynamic farming, this means understanding the farm not just as a physical entity but as a spiritual organism that interacts with the wider cosmos. Farmers are encouraged to develop a deep, intuitive understanding of their farm's needs and to make decisions based on both scientific principles and spiritual guidance.

1.2 The Farm as a Living Organism

A central tenet of biodynamic farming is the idea that the farm is a living organism, a self-contained and self-sustaining entity. This concept is a departure from conventional farming, which often views the farm as a factory for producing food, with inputs and outputs managed in isolation. In contrast, biodynamic farming sees the farm as an integrated whole, where every element is interconnected and plays a role in the overall health and balance of the system.

- **Holism:** In biodynamic farming, the whole is greater than the sum of its parts. The farm is seen as a living being with its own individuality, much like a person. This holistic approach means that the farmer's role is not just to manage resources but to nurture the farm's life forces. The health of the soil, plants, animals, and humans are all interconnected, and each influences the other.
- **Self-Sufficiency:** A biodynamic farm aims to be as self-sufficient as possible, minimizing external inputs. This self-sufficiency is achieved through practices like composting, crop rotation, and integrating livestock into the farming system. Animal manure is composted and returned to the soil as fertilizer, crop residues are used to build organic matter, and livestock grazing helps manage weeds and improve soil structure. The goal is to create a closed-loop system where waste is recycled, and resources are regenerated within the farm.
- **Biodiversity:** Biodiversity is a cornerstone of the biodynamic farm. A diverse range of crops, animals, and beneficial insects is encouraged to create a balanced and resilient ecosystem. This diversity helps prevent the spread of pests and diseases, enhances soil fertility, and supports a variety of wildlife. For example, integrating livestock with crop production can provide natural pest control, enhance nutrient cycling, and improve soil health.

- **The Role of the Farmer:** In the biodynamic view, the farmer is not just a laborer but a steward of the land, responsible for the well-being of the entire farm organism. This role requires a deep understanding of the farm's unique characteristics, including its soil, climate, plants, and animals. The farmer must be attuned to the farm's rhythms and needs, making decisions based on observation, intuition, and knowledge of biodynamic principles.

2. Preparing Your Farm for Biodynamic Practices

Preparing your farm for biodynamic practices involves several critical steps that ensure you lay a solid foundation for creating a self-sustaining and harmonious farming system. This preparation encompasses assessing your land, developing a comprehensive farm plan, and enhancing soil health—all essential for successfully implementing biodynamic principles.

2.1 Assessing Your Land

2.1.1 Evaluating Soil Health

The first step in preparing your farm is assessing the health of your soil. Soil is the foundation of any farming system, and in biodynamic farming, its health and vitality are paramount. Conduct a thorough soil analysis to understand its current condition, including:

- **Soil Composition:** Determine the soil's texture, including the proportions of sand, silt, and clay. This affects water retention, nutrient availability, and plant growth.
- **Soil pH:** Test the soil's pH level to ensure it is within the optimal range for your intended crops. Most plants prefer a slightly acidic to neutral pH (around 6.0 to 7.0).
- **Nutrient Levels:** Analyze the soil for essential nutrients like nitrogen, phosphorus, potassium, calcium, magnesium, and trace elements. This information helps guide your soil improvement efforts.
- **Organic Matter:** Measure the amount of organic matter in the soil. High levels of organic matter improve soil structure, moisture retention, and fertility.

Soil testing can be done through local agricultural extension services or private laboratories. Based on the results, you can develop a plan to address any deficiencies or imbalances.

2.1.2 Understanding Water Resources

Evaluate your farm's water resources, including the availability, quality, and distribution of water. Consider the following aspects:

- **Water Source:** Identify the sources of water for irrigation, such as wells, ponds, or rainwater harvesting systems. Ensure that these sources are sustainable and reliable.
- **Water Quality:** Test the water for contaminants, pH levels, and nutrient content. Clean, high-quality water is essential for plant health and composting.
- **Irrigation Systems:** Assess your current irrigation systems and their efficiency. Implement water-saving practices such as drip irrigation or rainwater collection to reduce water usage and environmental impact.

2.1.3 Climate and Weather Patterns

Understand the climate and weather patterns of your region, as they significantly impact farming practices. Key factors to consider include:

- **Temperature Ranges:** Know the average temperatures and seasonal variations. This information helps in selecting appropriate crops and planning planting and harvesting times.
- **Precipitation:** Analyze rainfall patterns and total annual precipitation. This affects soil moisture levels and irrigation needs.
- **Frost Dates:** Determine the average dates of the first and last frosts to plan planting and harvesting activities accordingly.

2.1.4 Existing Vegetation and Ecosystems

Assess the existing vegetation and ecosystems on your farm. This includes:

- **Current Vegetation:** Identify the types of plants and crops currently growing on the land. Understand their role in the ecosystem and how they interact with soil and climate conditions.
- **Wildlife and Beneficial Insects:** Observe the presence of wildlife and beneficial insects. These can provide natural pest control and contribute to biodiversity.

2.2 Developing a Farm Plan

2.2.1 Setting Goals and Objectives

Develop a clear vision for your farm and set specific goals and objectives. Consider:

- **Farm Size and Scope:** Define the scale of your farming operations, including the area of land to be cultivated and the types of crops and livestock to be raised.
- **Sustainability Goals:** Outline your sustainability goals, such as reducing chemical inputs, improving soil health, and enhancing biodiversity.

- **Economic Objectives:** Determine your economic goals, including target markets, revenue projections, and cost management.

2.2.2 Designing the Farm Layout

Create a detailed layout of your farm that incorporates biodynamic principles. This includes:

- **Field Layout:** Plan the arrangement of fields and plots based on crop rotation and companion planting principles. Designate areas for different crops, cover crops, and composting.
- **Livestock Integration:** Determine the location and housing for livestock. Integrate animals into the farm system to contribute to soil fertility and pest management.
- **Water Management:** Plan irrigation systems and water storage solutions to optimize water use and ensure efficient distribution.

2.2.3 Crop Rotation and Diversity

Design a crop rotation schedule that promotes soil health and reduces pest and disease pressures. Key considerations include:

- **Rotation Plans:** Rotate crops based on their nutrient needs and the benefits they provide to the soil. For example, follow nitrogen-fixing legumes with nutrient-demanding crops.
- **Diverse Plantings:** Incorporate a variety of crops and plant species to enhance biodiversity and resilience. Include cover crops and green manures to improve soil fertility and prevent erosion.

2.2.4 Implementing Biodynamic Preparations

Integrate biodynamic preparations into your farm plan to enhance soil fertility and plant health. This involves:

- **Preparation Production:** Learn how to produce biodynamic preparations, such as BD 500 and BD 501, using the appropriate materials and methods.
- **Application Schedule:** Develop a schedule for applying biodynamic preparations to soil, compost, and plants. Follow guidelines for timing and dilution to maximize their effectiveness.

2.3 Enhancing Soil Health

2.3.1 Building Organic Matter

Organic matter is crucial for soil health, as it improves soil structure, moisture retention, and nutrient availability. To build organic matter:

- **Composting:** Set up a composting system to recycle organic waste and produce nutrient-rich compost. Use biodynamic compost preparations to enhance compost quality and microbial activity.
- **Cover Crops:** Plant cover crops during fallow periods to add organic matter to the soil, prevent erosion, and improve nutrient cycling. Choose cover crops that suit your climate and soil conditions.

2.3.2 Improving Soil Structure

Healthy soil structure promotes root growth, water infiltration, and air circulation. To improve soil structure:

- **Minimal Tillage:** Practice minimal tillage to preserve soil structure and reduce erosion. Use techniques like no-till or reduced-till farming to maintain soil health.
- **Soil Amendments:** Apply organic soil amendments, such as compost, biochar, or green manures, to enhance soil structure and fertility.

2.3.3 Managing Soil Fertility

Maintain soil fertility by balancing nutrient levels and ensuring adequate availability for plant growth. Key practices include:

- **Nutrient Management:** Monitor soil nutrient levels and apply organic fertilizers as needed. Use compost, manure, or plant-based fertilizers to supply essential nutrients.
- **pH Adjustment:** Adjust soil pH if necessary to ensure it is within the optimal range for your crops. Use lime to raise pH or sulfur to lower pH, based on soil test results.

2.3.4 Implementing Biodynamic Practices

Incorporate biodynamic practices to enhance soil health and overall farm vitality. These practices include:

- **Biodynamic Calendar:** Use the biodynamic planting calendar to align your farming activities with cosmic rhythms. Follow the calendar for optimal planting, cultivating, and harvesting times.
- **Planting and Cultivation:** Apply biodynamic principles to planting and cultivation activities. This includes planting in harmony with lunar cycles and using biodynamic preparations to support soil and plant health.

3. Implementing Biodynamic Farming Practices

Implementing biodynamic farming practices involves a holistic approach to agriculture that integrates spiritual, ecological, and practical principles.

Biodynamic farming goes beyond traditional organic methods by emphasizing the farm as a self-sustaining ecosystem, guided by cosmic rhythms and supported by specific preparations. This section elaborates on the core practices of biodynamic farming, including soil management, crop management, animal husbandry, and the use of biodynamic preparations.

3.1 Soil Management

3.1.1. Soil Fertility and Health

Soil fertility is the cornerstone of biodynamic farming, and maintaining soil health is critical for a thriving farm. Biodynamic practices focus on building and sustaining soil fertility through natural processes:

- **Composting:** Composting is essential in biodynamic farming for recycling organic matter and enhancing soil fertility. Biodynamic compost includes preparations like BD 502 (Yarrow), BD 503 (Chamomile), BD 504 (Stinging Nettle), and BD 505 (Oak Bark), which are added to compost piles to stimulate microbial activity and improve compost quality. Compost should be well-aerated and turned regularly to ensure decomposition and the development of beneficial microorganisms.
- **Green Manures and Cover Crops:** Green manures and cover crops are planted to improve soil structure, prevent erosion, and enhance nutrient availability. They add organic matter to the soil, support beneficial microorganisms, and help in nitrogen fixation. For example, legumes like clover and vetch are commonly used as green manures due to their ability to fix atmospheric nitrogen.
- **Biodynamic Preparations:** Biodynamic preparations are used to enhance soil vitality and promote a healthy soil ecosystem. Key preparations include BD 500 (Horn Manure), which improves soil structure and fertility, and BD 501 (Horn Silica), which enhances photosynthesis and plant resilience. These preparations are applied at specific times according to the biodynamic calendar to align with cosmic rhythms.

3.1.2 Soil Structure and Erosion Control

Maintaining soil structure and preventing erosion are crucial for long-term soil health:

- **Minimal Tillage:** Biodynamic farming emphasizes minimal tillage to preserve soil structure and reduce erosion. Excessive tillage can disrupt soil life, deplete organic matter, and lead to compaction. Instead, adopt practices such as no-till or reduced-till farming to maintain soil integrity.

- **Erosion Control Measures:** Implement measures to control soil erosion, such as planting cover crops on slopes, building terracing, and using mulch to protect soil surfaces. Mulch helps retain moisture, regulate soil temperature, and reduce erosion.

3.2 Crop Management

3.2.1 Crop Rotation and Diversity

Crop rotation and diversity are integral to biodynamic farming practices, promoting soil health and reducing pest and disease pressures:

- **Crop Rotation:** Develop a crop rotation plan to avoid depleting soil nutrients and disrupting pest cycles. Rotate crops based on their nutrient requirements and the benefits they provide to the soil. For example, follow heavy feeders like corn with nitrogen-fixing legumes such as beans or peas.
- **Diverse Plantings:** Incorporate a variety of crops and plant species to enhance biodiversity and resilience. This includes growing different types of vegetables, fruits, grains, and herbs. Biodiversity helps prevent the spread of pests and diseases and supports beneficial insects and wildlife.

3.2.2 Pest and Disease Management

Pest and disease management in biodynamic farming focuses on natural and preventative methods:

- **Beneficial Insects:** Encourage beneficial insects that prey on pests, such as ladybugs, lacewings, and predatory beetles. Provide habitats for these insects through diverse plantings and creating insect-friendly environments.
- **Biodynamic Preparations for Pest Control:** Use biodynamic preparations to manage pests and diseases. For example, BD 507 (Valerian) is used to strengthen plant defenses and repel pests. Preparation BD 508 (Dandelion) can enhance plant resilience and growth.
- **Companion Planting:** Implement companion planting strategies to deter pests and enhance plant growth. For example, planting marigolds near vegetables can repel nematodes and other pests.

3.2.3 Planting and Harvesting

Aligning planting and harvesting activities with cosmic rhythms is a key aspect of biodynamic farming:

- **Biodynamic Calendar:** Follow the biodynamic planting calendar to determine the optimal times for planting, cultivating, and harvesting based on lunar and planetary cycles. The calendar divides days into periods favorable for root, leaf, flower, or fruit growth, guiding farmers in their agricultural activities.

- **Planting Techniques:** Use planting techniques that promote strong plant growth and vitality. This includes proper spacing, planting depth, and soil preparation. Ensure that plants are healthy and well-established before the onset of adverse weather conditions.

3.3 Animal Husbandry

3.3.1 Integrating Livestock into the Farm System

Integrating livestock into the farm system is a key practice in biodynamic farming, enhancing soil fertility and farm productivity:

- **Livestock Management:** Choose appropriate breeds and manage livestock to support soil health and farm productivity. Livestock such as cows, chickens, and pigs can contribute to soil fertility through manure production and grazing.
- **Manure Management:** Use animal manure as a valuable resource for soil fertility. Compost manure to create nutrient-rich compost and apply it to fields to improve soil health. Biodynamic preparations like BD 500 (Horn Manure) enhance the effectiveness of manure applications.
- **Grazing Systems:** Implement rotational grazing systems to optimize pasture use and prevent overgrazing. This approach helps maintain healthy pastureland, improves soil structure, and supports diverse plant species.

3.3.2 Animal Welfare

Ensure high standards of animal welfare by providing adequate housing, nutrition, and healthcare:

- **Housing and Space:** Provide comfortable and spacious housing for livestock, allowing them to exhibit natural behaviors and access clean water and feed.
- **Nutrition and Health:** Feed livestock a balanced diet that meets their nutritional needs and monitor their health regularly. Implement preventative measures to manage diseases and parasites.

3.4 Biodynamic Preparations and Applications

3.4.1 Preparation of Biodynamic Preparations

Biodynamic preparations are integral to enhancing farm vitality and promoting a balanced ecosystem:

- **Preparation Techniques:** Follow precise techniques for preparing biodynamic preparations, such as BD 500 (Horn Manure) and BD 501 (Horn Silica). For BD 500, fill cow horns with fresh manure and bury them in the soil over the winter months. For BD 501, fill cow horns with ground quartz and bury them over the summer. These preparations must be carefully processed and applied to the farm.
- **Application Methods:** Apply biodynamic preparations at specific times according to the biodynamic calendar. Dilute the preparations in water and spray them on soil, compost, or plants. Proper application ensures that the preparations enhance soil and plant health effectively.

3.4.2 Timing and Alignment

The timing of biodynamic preparation applications is based on lunar and planetary cycles:

- **Lunar Cycles:** Use the biodynamic calendar to determine the optimal days for application. The calendar divides days into periods favorable for different types of plant growth (root, leaf, flower, fruit) and aligns with the lunar phases.
- **Planetary Influences:** Consider planetary influences, such as the positions of planets and stars, when planning applications. Certain planetary alignments are believed to enhance the effectiveness of biodynamic preparations.

3.5 Monitoring and Adjusting Practices

3.5.1 Observing Farm Conditions

Regular observation of farm conditions is essential for adapting and refining biodynamic practices:

- **Farm Monitoring:** Monitor soil health, plant growth, and livestock behavior regularly. Use observations to identify issues and make informed decisions about adjustments to practices.
- **Record Keeping:** Maintain detailed records of farm activities, including planting dates, preparation applications, and crop yields. This information helps track the effectiveness of biodynamic practices and informs future decisions.

3.5.2 Adapting to Challenges

Adapt biodynamic practices to address challenges and optimize farm performance:

- **Problem Solving:** Address issues such as pest infestations, soil imbalances, or weather-related challenges using biodynamic principles. Seek advice from experienced biodynamic farmers and consult resources like biodynamic associations and research publications.
- **Continuous Improvement:** Continuously improve your farming practices by experimenting with new techniques, learning from successes and failures, and staying informed about advancements in biodynamic farming.

4. Aligning with the Biodynamic Calendar

4.1 Understanding Cosmic Rhythms

One of the unique aspects of biodynamic farming is the alignment of agricultural activities with cosmic rhythms. The biodynamic calendar, which is based on lunar and planetary cycles, guides farmers in timing their planting, cultivating, and harvesting activities.

The calendar divides the year into four main periods, each associated with one of the classical elements—earth, water, air, and fire. These periods are further subdivided into days that are best suited for specific tasks, such as planting root crops, pruning fruit trees, or applying compost.

4.2 Using the Biodynamic Calendar

To use the biodynamic calendar effectively, familiarize yourself with its structure and how it relates to your specific crops and farming activities. For example, root crops like carrots and potatoes are best planted during earth days, while leafy greens thrive when planted on water days.

The biodynamic calendar also provides guidance on the optimal times for applying biodynamic preparations and conducting other farm tasks. By aligning your activities with these cosmic rhythms, you can enhance the vitality of your crops and create a more harmonious farm environment.

5. Certification and Market Opportunities

5.1 Biodynamic Certification

Biodynamic certification is a way to demonstrate your commitment to the principles and practices of biodynamic farming. The certification process is overseen by organizations such as Demeter International, which sets strict standards for biodynamic farming practices.

To become certified, your farm must meet specific criteria, including the use of biodynamic preparations, crop rotation, composting, and adherence to the biodynamic calendar. The certification process typically involves an on-site inspection, a review of farm records, and ongoing compliance with certification standards.

5.2 Market Opportunities

Biodynamic products are increasingly recognized for their high quality and sustainability. As consumer demand for organic and sustainably produced foods grows, biodynamic farmers have access to premium markets that value these attributes. Certified biodynamic products often command higher prices and are sought after by health-conscious consumers, chefs, and retailers.

In addition to direct marketing their produce through farmers' markets, community-supported agriculture (CSA) programs, and farm stands, biodynamic farmers can explore opportunities in specialty grocery stores, organic food cooperatives, and restaurants that prioritize sustainable sourcing. The rising awareness and appreciation of biodynamic farming's principles and benefits offer a unique branding opportunity for farmers to distinguish their products in a competitive market.

5.3 Building a Community and Network

Biodynamic farming is often about more than just growing food; it's about building a community of like-minded individuals who are committed to sustainable and regenerative agriculture. Joining local and international biodynamic farming associations can provide valuable support, education, and networking opportunities. These organizations often host workshops, conferences, and farm tours where you can learn from experienced farmers, share your challenges, and find inspiration.

Engaging with your local community through educational events, farm visits, and volunteer programs can also help raise awareness about biodynamic farming and its benefits. By fostering a strong community around your farm, you can build lasting relationships with customers, neighbors, and fellow farmers who share your values.

6. The Journey of Continuous Learning

6.1 Adapting to Challenges

Biodynamic farming is a dynamic and evolving practice that requires continuous learning and adaptation. Each farm is unique, and what works well in one location may not be suitable for another. As a biodynamic farmer, you must be prepared to observe, experiment, and adjust your practices based on the specific conditions of your land and the needs of your farm.

Challenges such as climate change, soil degradation, and market fluctuations can impact your farming practices. However, the principles of biodynamic farming—such as building soil health, promoting biodiversity, and aligning with natural rhythms—provide a resilient foundation that can help you navigate these challenges and thrive.

6.2 Staying Informed

Staying informed about the latest research, trends, and innovations in biodynamic farming is essential for success. Subscribe to agricultural journals, attend conferences, and participate in online forums to keep up with new developments and share your experiences with others in the biodynamic community.

Learning from the successes and failures of other biodynamic farmers can also provide valuable insights. Case studies, farm tours, and mentorship programs can offer practical guidance and inspire new ideas for improving your farm.

Conclusion

Getting started with biodynamic farming is a journey that requires a deep commitment to sustainability, a willingness to learn, and a connection to the land. By understanding the philosophical foundations of biodynamic farming and implementing practical techniques like soil health improvement, crop rotation, composting, and pest management, you can create a self-sustaining farm ecosystem that nurtures both the environment and your community.

As you embark on this path, remember that biodynamic farming is not just about following a set of rules; it's about cultivating a holistic and regenerative approach to agriculture that honors the interconnectedness of all living things. Whether you're growing food for your family, your community, or a broader market, biodynamic farming offers a rewarding and meaningful way to contribute to a healthier planet.

References

Koepf, H.H. (2005). The Biodynamic Farm: Agriculture in Service of the Earth and Humanity. SteinerBooks.

Olsen, E. K. (2014). Biodynamic agriculture: a valuable alternative to the industrial farming system.

Phillips, J.C., & Rodgriguez, L. P. (2006). Beyond organic: an overview of biodynamic agriculture with case examples.

Steiner, R. (2004). Agriculture course: the birth of the biodynamic method. Rudolf Steiner Press.

Van Kraalingen, I. (2019). Cultivating embodied connections in biodynamic agriculture (Master›s thesis).

Index

For Product Safety Concerns and Information please contact our EU representative GPSR@taylorandfrancis.com Taylor & Francis Verlag GmbH, Kaufingerstraße 24, 80331 München, Germany